EXPERIMENTAL RESEARCH

Experimental Research: Brief Essays on Timeless Concepts offers a unique exploration of the experimental process that transcends disciplinary boundaries and technological trends. Through a series of concise, thought-provoking essays, this book distills enduring principles of scientific inquiry. Addressing foundational topics from measurement and dimensional analysis to cognitive factors in research and technical communication, it cultivates a deep, reflective understanding of experimentation as both an intellectual craft and a scientific discipline.

Organized into seven thematic sections, the essays delve into topics rarely addressed in conventional manuals, including the epistemology of experimentation, creative thinking in research, principles of device design and the thoughtful integration of modern tools into contemporary experimental work. Rather than focusing on discipline-specific techniques, the authors emphasize universal aspects of experimentation that cut across fields, blending conceptual insight with practical relevance and historical perspective. The result is a timeless perspective on experimental methods that balances accessibility with intellectual rigor.

This volume is an essential resource for young scholars learning to design, conduct and interpret experiments independently, such as graduate students, postdoctoral researchers and early-career faculty who, despite solid technical backgrounds, have received little formal guidance in the conceptual and cognitive aspects of experimentation. For educators and established researchers, it offers fresh perspectives on mentoring the next generation of experimentalists in an era of rapid technological change.

J. Carlos Santamarina is Professor and Clough Chair at the Georgia Institute of Technology (United States). He earned his degrees from the Universidad Nacional de Córdoba (Argentina), the University of Maryland and Purdue University (United States) and has held academic appointments at NYU-Poly (United States), the University of Waterloo (Canada), KAUST (Saudi Arabia) and Georgia Tech. His research integrates experimental and numerical approaches to investigate complex phenomena in energy geotechnics. A frequent invited speaker, Santamarina has delivered many of the field's landmark lectures and has received numerous international honors and awards.

Andrea Muñoz-Ibáñez is a Ramón y Cajal Fellow at the Universidade da Coruña (Spain) and a former Marie Curie Postdoctoral Fellow. She earned her Ph.D. from the same university and has held postdoctoral positions at the Universidade de Vigo (Spain), KAUST (Saudi Arabia) and Georgia Tech. Her experimentation-centered research integrates experimental and numerical approaches to explore materials and coupled multiphysics processes under extreme conditions. Supported by national and international funding agencies, she brings perspectives rooted in the current realities and cultural trends shaping the next generation of researchers.

EXPERIMENTAL RESEARCH

Brief Essays on Timeless Concepts

J. Carlos Santamarina and Andrea Muñoz-Ibáñez

Routledge
Taylor & Francis Group
LONDON AND NEW YORK

Designed cover image: Getty Images

First published 2027
by Routledge
4 Park Square, Milton Park, Abingdon, Oxon OX14 4RN

and by Routledge
605 Third Avenue, New York, NY 10158

Routledge is an imprint of the Taylor & Francis Group, an informa business

British Library Cataloguing-in-Publication Data
A catalogue record for this book is available from the British Library

ISBN: 978-1-041-21494-6 (hbk)
ISBN: 978-1-041-20311-7 (pbk)
ISBN: 978-1-003-72752-1 (ebk)

DOI: 10.4324/9781003727521

Typeset in Sabon
by SPi Technologies India Pvt Ltd (Straive)

To our families

CONTENTS

PREFACE

The development of logic, mathematical analysis, numerical methods and recent advances in artificial intelligence have prompted a re-evaluation of the role of experimentation in science and engineering. While new modes of knowledge creation inevitably reshape established approaches, these developments have ultimately reinforced the foundational importance of experimental research.

The essays in this volume focus not on fleeting technological trends or the latest gadgets, which quickly become outdated, but on enduring concepts and principles. They are intended for readers who seek understanding through careful experimentation, guided by curiosity, patience and thoughtful rigor.

Each essay is concise, distilling key lessons and best practices in experimental research. Many topics could justify entire books, and every statement invites deeper exploration. Our goal is to introduce readers to the mindset and methods of experimental research, fostering attention to physical, cognitive and analytical understanding. While the content applies broadly across disciplines, essays on experimental devices will speak more directly to researchers in the physical sciences and engineering.

This book is complemented by a dynamic and evolving website, which provides supporting examples, additional exercises and project ideas that expand upon many of the themes introduced here.

We open each essay with a quote, not to invoke authority, but to capture a central idea in a brief, often witty line. The insights that follow draw on our own experiences, shaped by the collective knowledge of thinkers and scientists throughout history. Omissions are inevitable in a work of this scope, and some ideas may trace back to sources we no longer recall.

Artificial intelligence tools have become widely accessible in recent years and now touch many aspects of research, including writing. While we wrote all the essays and sketched all the figures in this book, we used AI-based tools to assist with editing the final versions of the text and selected images.

We thank the many reviewers for their valuable feedback: G. Abelskamp, N.F. Bolson, W. Cai, A. Cardona, G.M. Castro, J. Delgado, D. Fratta, A. Hafez, G. Han, S. Jung, R.M. Rached, M.C. Santamarina, M. Terzariol and J.Y. Yeo. We also thank our colleagues, who have shared their knowledge so generously over the years, as well as the funding agencies whose support made it possible to build and sustain our research programs. Above all, we are most grateful to our families for their continued encouragement.

We suggest reading this manuscript alongside an ongoing experimental project, whether in a classroom or professional setting. Essays may be read in any order but do begin with the essay on safety (Essay 38). Experimentation continues to fuel vibrant intellectual lives. Enjoy this fascinating journey!

SECTION 1
Preliminary concepts

Essay 1

KNOWLEDGE CREATION THROUGHOUT HISTORY

Nature provides extraordinary pleasures for those who can recognize the causes in things[1]

How did we come to know what we know? Today's body of knowledge is the result of countless cumulative insights, built over centuries through observation, reasoning and refinement. This long journey has been shaped by steady incremental advancements and, at times, by profound paradigm shifts, such as N. Copernicus's (1473–1543) heliocentrism or C. Darwin's (1809–1882) theory of evolution. Throughout history, the methods of discovery have evolved – empirical, rational, analytical and computational – often propelled by technological innovation and driven by visionary individuals whose ideas reshaped entire disciplines. Yet even as our tools advanced, some of the most powerful means of discovery have remained simple and timeless. All driven by the same impulse: to understand the causes and patterns behind the world around us.

Empiricism

Learning through observation is as old as humanity itself. Early communities learned to interpret weather patterns, animal behavior and the properties of materials through sensory experience. That same empirical foundation underlies today's *scientific method*[2] and tacitly supports most engineering practice. The Romans (27 BCE–1453 CE) relied on *trial-and-error* to understand the interplay between material properties and geometric shapes,[3] culminating in remarkable structures like the Pantheon in Rome, still the world's largest unreinforced concrete dome. T. Edison (1847–1931) tested thousands of materials in search of a durable light-bulb filament. In nature, trial-and-error

DOI: 10.4324/9781003727521-2

combined with natural selection drives Darwinian evolution and today inspires genetic algorithms used for search and optimization. Yet experience alone cannot always reveal the underlying causes.

Rationalism

Socrates (470 BC–399 BC) and Plato (428 BC–348 BC) sought understanding through *sequential questioning, reflection and logical analysis*. Aristotle (384 BC–322 BC) integrated *qualitative observations and hypothetical reasoning* to explore natural phenomena. Centuries later, E. Mach (1838–1916) and A. Einstein (1879–1955) used *thought experiments* and "what-if" scenarios to develop new theories when physical experiments were not feasible. Einstein's theory of relativity (1905) emerged from such a reflective approach and was later confirmed by A. Eddington and F. Dyson during the solar eclipse of 1919.

Mathematical analysis

Rational reflection found its most powerful ally in mathematics, which transformed intuition into quantifiable prediction. G. Galilei (1564–1642) played a central role in the scientific revolution by bringing together *systematic quantitative experimentation and mathematical formulation*. He improved the telescope, made extensive celestial observations, verified Copernicus's heliocentric model and captured his experimental observations in mathematical terms. He famously claimed that the grand book of the universe "is written in the language of mathematics".

The development of *mathematical analysis* marked a turning point in knowledge creation. Calculus, independently advanced by G.W. Leibniz (1646–1716) and I. Newton (1643–1727), sparked rapid growth in engineering and science. Notably, J.C. Maxwell (1831–1879) synthesized earlier contributions from C.A. de Coulomb, H.C. Ørsted, A.M. Ampère, M. Faraday and C.F. Gauss into a unified mathematical framework that brought together electricity, magnetism and optics[4]; and predicted electromagnetic waves, which were confirmed by H. Hertz (1857–1894) two decades later. Since then, mathematical reasoning has expanded beyond physics and engineering to model complex systems in biology, economics and even human behavior.

Computer-based exploration

Since the 1940s, computers have redefined how we generate knowledge. *Numerical modeling* now spans an immense range of scales, from atomistic studies to Earth-scale analyses, covering 17 orders of magnitude in length and more than 28 orders of magnitude in time! Its impact is felt across all fields, from high-energy physics and atmospheric science to tumor modeling and aircraft design.

Once considered an oxymoron, *numerical experiments* now play a vital role in testing hypotheses, simulating complex systems and exploring conditions beyond the reach of physical experimentation. Meanwhile, advancements in sensors, microelectronics, computing and communications have given rise to *IT-augmented experimentation* and *digital twins*. These technologies allow for refined experimental designs, intricate real-time boundary control,[5] concurrent back-analysis of multi-physics data streams using complex numerical simulators and the upscaling of experimental results to real-world conditions.[6]

More recently, widespread data collection and advances in information technology have produced massive databases. Improvements in statistical methods and machine learning algorithms have empowered researchers to analyze these complex *datasets*, uncover patterns, test hypotheses and advance scientific understanding. Reflecting the *transformative impact of AI* on knowledge creation, the 2024 Nobel Prize in Physics was awarded jointly to J. Hopfield and G. Hinton for their foundational contributions to neural networks and machine learning. In the same year, the Nobel Prize in Chemistry was awarded to D. Hassabis, J.M. Jumper and D. Baker, in part for AI-based protein structure prediction and computational protein design.

Back to basics

Despite our expanding toolset, the essentials of knowledge creation remain unchanged: *curiosity and deep, clear thinking*. Moreover, some tools may go unnoticed in today's high-tech world yet remain deeply embedded in how we build understanding. A *mindful walk* with a collaborator fosters creativity, opens space for alternative perspectives, strengthens ideas through argumentative dialogue[7] and promotes clarity by requiring concepts to be articulated aloud.

Describing, comparing, clustering and *systematically organizing information* help identify essential characteristics and governing processes. The periodic table of elements is a prime example: D.I. Mendeleev systematically ordered the known elements based on the available information at the time (1869); although only 63 of the 118 elements were known, he left gaps for the unknown ones and even predicted some of their properties.

The act of *drawing and sketching* disciplines attention,[8] sharpens observation, aids in understanding complex processes and enables us to visualize order in information. L. da Vinci (1452–1519) used sequential sketches to test hypotheses and progressively understand the mechanics of the human body; C. Darwin (1809–1882) visualized evolution through his tree of life; J.J. Audubon (1785–1851) captured nature in meticulous illustrations of birds[9]; and R. Feynman (1918–1988) translated complex mathematical expressions into intuitive diagrams that revealed interactions among subatomic particles. The same visual logic underlies concept maps and flowcharts used across all modern disciplines.

Finally, *scientific writing* involves assembling a puzzle of concepts, critically examining misaligned ideas and identifying missing pieces until everything fits into a coherent narrative that integrates prior knowledge with new findings. Writing is therefore not merely a means of reporting and communication, but a powerful instrument of discovery and knowledge creation.

A tango dance

The emergence of new tools for knowledge creation has transformed the role of empirical observation and experimentation in research. However, experiments retain their intrinsic value by capturing the full complexity of physical reality, rather than relying on analytical or numerical abstractions. In fact, there is renewed interest in experimental methods, fueled in part by the increasing availability of powerful and affordable instruments. This democratization of technology broadens the scope of experimental research, fosters creative approaches and enables greater participation and collaboration. The possibilities ahead are truly exciting!

Discovery is a "tango dance" between experimentation and other forms of knowledge creation, sometimes moving in opposition yet always intimately intertwined in the pursuit of understanding, the search for solutions and the advancement of knowledge.

Exercises

1 Identify great scientists in your field of interest (or beyond). Explore their lives and main contributions.
2 Explore different methods of knowledge creation and determine which ones align best with your field of research and your approach.

3 The disruptive power of numerical methods and artificial intelligence offers new opportunities to enhance the creative and interpretative dimensions of scientific inquiry. How could these tools be integrated into your own research? In what ways might they influence the questions you ask or the approaches you take?
4 Analyze and discuss: If you can simulate the problem under study, why would you still conduct experiments? What factors might lead to unexpected outcomes?
5 Explore the power of drawing as a means of discovery. Visualize a bicycle without looking at one directly; sketch its essential components. Then iterate on your sketches, gradually refining them until you develop a mechanically feasible and structurally sound design.

Notes

1 Aristotle (384 BC–322 BC), a Greek philosopher, explored logic, ethics, physics and biology.
2 See essay on *Hidden Magic in Science and Engineering (Section 2).*
3 See essay on *Material, Form and Function (Section 4).*
4 See essay on *Our Physical Universe: Laws and Principles (Section 1).*
5 The antilock braking system, popularized since the 1970s, is an example of a sensor-computer-actuator system for real-time boundary control.
6 The use of "digital twins" is discussed further in the essay on *Devices: Design Strategies Enhanced by Digital Twins* (*Section* 4).
7 *Socratic Method.*
8 S. Ramón y Cajal (1897). *Advice for a Young Investigator*. MIT Press.
9 J.J. Audubon, *The Birds of America*, sections published between 1827 and 1838.

Essay 2

OUR PHYSICAL UNIVERSE

Laws and principles

Truth is ever to be found in simplicity, and not in the multiplicity and confusion of things[1]

We have studied so many theories and equations! It is easy to forget that every phenomenon we observe, no matter how complex, ultimately stems from a small set of universal laws. Let us take a moment to recall these fundamental laws and a few guiding principles that underpin the physical universe. *Let's begin at the beginning.*[2]

The fundamental laws

Laws of Motion (17th century). Newton's three laws describe how forces affect motion: (1) in the absence of a net force, a body remains at rest or moves with constant velocity; (2) a net force F produces an acceleration a proportional to the mass m, $F = ma$; and (3) two interacting bodies experience equal but opposite forces. Newton's gravitational force $F = G\frac{m_1 m_2}{r^2}$, which relates two masses m_1 and m_2 separated by a distance r, through the gravitational constant G, obeys the basic laws; while it was later superseded by general relativity, it remains remarkably accurate for most practical applications.

Electromagnetism (19th century). Maxwell's equations unify electricity and magnetism: (1) the surface integral of the electric field equals the charge enclosed by the surface; (2) the surface integral of the magnetic flux through a closed surface is zero since magnetic poles cannot be isolated; (3) a time-varying magnetic field induces an electric field; and (4) electric currents and

DOI: 10.4324/9781003727521-3

time-varying electric fields produce magnetic fields. Together, these equations predict electromagnetic waves, including light. The associated Coulomb's law $F = \frac{1}{4\pi\varepsilon_0}\frac{q_1 q_2}{r^2}$ describes the electrostatic force between two stationary charges q_1 and q_2 separated by a distance r, where ε_0 is the vacuum permittivity.

Thermodynamics (19th century). These laws govern energy flow in all systems. The first law prescribes that the total *energy in an isolated system is conserved*; thus, energy is neither created nor destroyed, only transformed. The second law introduces entropy as a measure of disorder and states that the *entropy of an isolated system always increases* with time, providing a criterion for the *spontaneity* of natural processes. The third law states that the entropy of a perfect crystal approaches zero as the temperature approaches absolute zero, establishing an absolute reference for entropy.

Conservation Laws. Certain physical quantities remain constant within a closed system, regardless of the transformations taking place: *energy* (first law of thermodynamics), *mass*, *linear and angular momentum* and *electric charge.*[3] Conservation laws provide the foundation for analyzing and predicting complex phenomena. For example, steady-state Laplacian flow and Fickian diffusion analyses start with mass conservation,[4] Kirchhoff's equations assume charge conservation, and collision problems invoke momentum conservation.

Extensions. While these classical laws govern most of our physical reality, developments in the 20th century extended physics into more extreme domains. *Relativity* describes phenomena that occur near the speed of light and in strong gravitational fields, as observed in particle accelerators and near black holes. *Quantum mechanics*, in turn, explains the behavior of matter and energy at very small scales, including atoms, electrons and subatomic particles.

Guiding principles

Several overarching principles guide the study of natural phenomena.

Le Chatelier's Principle. Nature resists change: when a system at equilibrium is perturbed, it adjusts to counteract the imposed change.

Ockham's Razor. When multiple explanations are possible, the one that accounts for the evidence with the fewest assumptions is usually preferred. Parsimony is a hallmark of effective scientific theories.

Beauty. Many physical theories exhibit unique "elegance", often reflected in their internal consistency, symmetry and simplicity. Beautiful theories typically have broad relevance and predictive power and may even reveal unexpected connections between otherwise distinct observations.

Final insight

Recognizing the connections between physical laws and guiding principles can open new ways of understanding the world. Consider the following example: stretch a rubber band. Part of the strain energy is transformed into thermal energy, barely noticeable yet detectable if you touch the band with your lips. Keep it stretched until the heat dissipates, then release it quickly – it will feel cooler. This experiment reveals the interplay between Le Chatelier's principle and energy conservation.

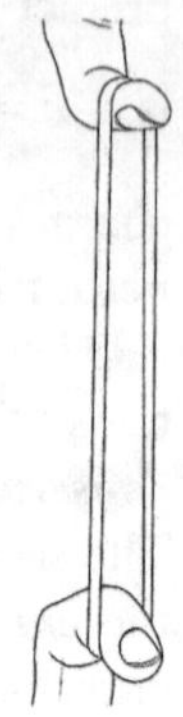

Such interactions hint at the broader coupling among different forms of energy: mechanical (kinetic, potential, strain energy), chemical, nuclear, electromagnetic (radiant), thermal and acoustic. Energy coupling underlies most sensors, from biological systems to microelectronics: photoreceptors in our eyes convert light into neural impulses, while piezoelectric elements transform mechanical strain into electrical signals.

A deep understanding of these foundational laws and principles guides thoughtful research, inspires creative experimental designs and invites us to appreciate the subtle harmony inherent in physical systems.

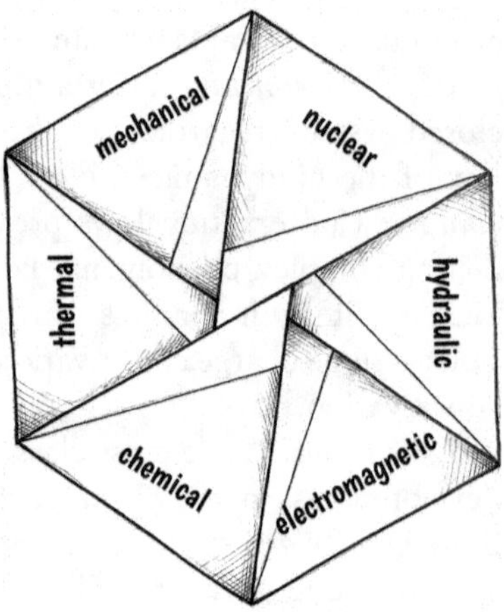

Exercises

1 Provide specific examples from your field for each fundamental physical law.
2 List and describe instances of energy coupling in your field.
3 Compare competing explanations of a phenomenon that interests you. Evaluate them in terms of both Ockham's parsimony and accuracy.
4. Three models are fitted to the same dataset. Which one would you choose, and why?

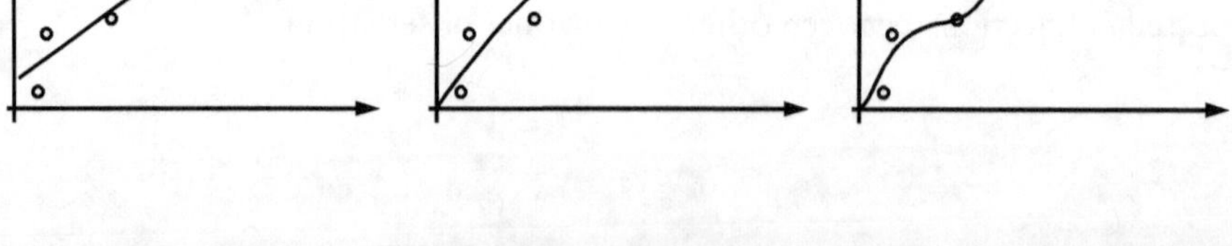

Notes

1 I. Newton (1643–1727), English mathematician, physicist and astronomer.
2 L. Carroll (1832–1898), English writer, mathematician, logician and photographer.
3 Additional conservation laws extend to the relativistic and quantum domains.
4 See essay on *Differential Equations: Processes in Mathematical Form (Unstated Assumptions) (Section 5).*

Essay 3

A GREAT QUESTION

Where good science begins

The man who asks a question is a fool for a minute, the man who does not ask is a fool for life[1]

Research fields do not lose momentum because all their questions have been answered. They fade because their questions stop being relevant. A field remains vibrant only as long as it continues to ask questions that matter, spark curiosity and resonate with society's evolving priorities.

Our questions reveal the depth of our understanding, the clarity of our thought and our ability to notice what others may overlook. Ultimately, they define the impact of our work.

Good questions do not appear by chance. Asking them requires sustained curiosity, focused thought and the courage to think differently. They emerge when we step away from familiar routines, engage with others and learn how different fields approach similar challenges. Above all, they require making time to think.

Thinking beyond the obvious

Interesting

The set of possible questions is infinite. Most are easy, and some are even lazy. Questions that begin with "Is it?" or "What is?" can often be answered quickly and superficially. They rarely open new doors.

Instead, seek *interesting questions* that challenge assumptions and invite deeper understanding. They do not have to be complicated. Often, the most revealing questions arise from re-examining what we take for granted and

DOI: 10.4324/9781003727521-4

the obvious: Why do clouds remain high in the sky when water vapor is heavier than air? Why do tree branches divide in characteristic patterns? Some of the most fertile ideas emerge when we resist intellectual crowds and pursue orthogonal lines of research.[2]

Impactful

Not every interesting question is worth pursuing. Some are clever but inconsequential. Institutions often reward safe and incremental work, yet we must ask ourselves why we would spend years or even an entire career chasing answers that do not matter. Time and intellectual energy are limited resources. Choose questions that challenge your thinking and move knowledge forward.

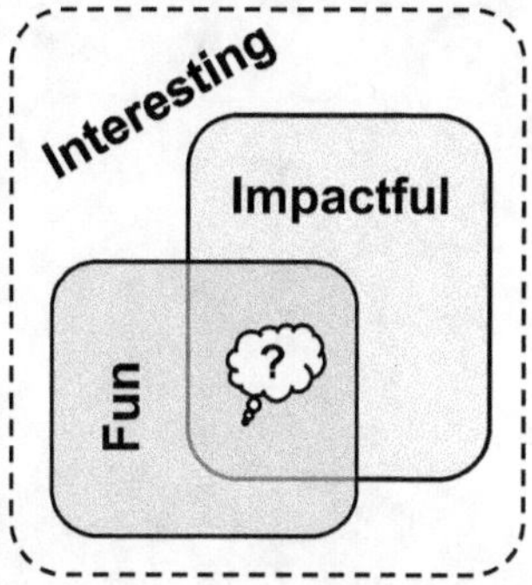

The best questions are both *interesting and consequential*. Their answers, if found, could change the way we see the world. Of course, relevance is not always clear in advance. When J.D. Watson and F.H.C. Crick set out to determine the structure of DNA, they could not have foreseen the full scope of its implications, but they understood that the problem was fundamental to biology and life itself.

And fun...

Joy is another essential criterion. Experimental research demands time, patience and perseverance. The most enduring motivation comes not from duty or ambition, but from curiosity and genuine excitement.

Ask, discover, thrive

Young researchers operate under pressure, continuously balancing both institutional and personal demands. Yet, it is crucial to make time to think deliberately and persistently, to identify those questions that are interesting, impactful and fun.

By cultivating curiosity, pursuing meaningful questions and enjoying discovery, you will not only shape your own career but also help ensure that your field continues to thrive.

Exercises

1 Develop a clear strategy to improve how you ask research questions.
2 Identify engaging questions from everyday observations, for example, in the kitchen or from our senses.

3 Describe your favorite research topic. Generate seven questions. Select the most meaningful and fun ones.
4 Analyze and discuss AI's ability to ask versus answer questions.

Notes

1 Proverb often attributed to Confucius (551–479 BCE), Chinese philosopher.
2 For a mathematical metaphor, consider the conjugate gradient method to identify the function minimum instead of the steepest descent.

Essay 4

LITERATURE REVIEW

Mapping the knowledge landscape

أرقإ = *Read*[1]

What's past is prologue[2]

Our lives are surrounded by books, databases, the infinite web… so much information! Purposeful engagement with this vast body of work, coupled with healthy skepticism and openness to alternative views, is the foundation for effective and impactful research.

Goals for a great review

Identify the boundary of knowledge

We operate within the space of the known and strive to push its limits. The chance of suddenly landing at the edge of knowledge without preparation is vanishingly small. A literature review maps this boundary, revealing what is known and highlighting gaps that guide meaningful research questions. A good mentor and a strong team can help position young researchers to see into the unknown.[3]

One early warning sign of a weak review is poor references. While reviewing recent work is essential, revisiting the classics is equally important: much of modern knowledge rests on a small number of foundational contributions. In fact, it has been argued that 90% of scientific knowledge has been produced by just 10% of investigators.[4]

DOI: 10.4324/9781003727521-5

Explore broadly and across disciplines[5]

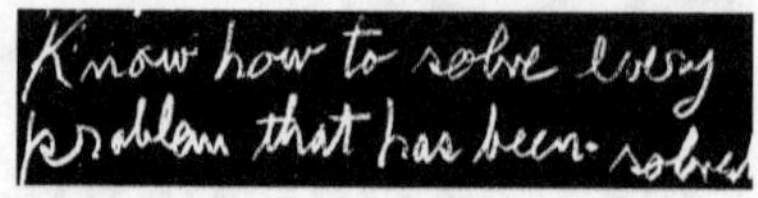

Knowledge knows no boundaries. Limiting your search to your immediate field risks missing key insights. Explore analogous problems across disciplines; the question you are asking may already have been addressed elsewhere. Engaging with these solutions sharpens your thinking and exposes assumptions often taken for granted. In some cases, this exploration may reveal a hidden pathway to breakthroughs; for example, Einstein challenged Newton's concept of absolute space and time, leading to the development of the theory of relativity. Broad exploration also increases the relevance and impact of your work across domains.

Gain perspective - see further

The higher we climb upon the heap of knowledge, the further we can see beyond its horizon, and the more pointed and far-reaching our questions become.[6] In Newton's words, *if I have seen further, it is by standing on the shoulders of giants.*[7]

Synthesize information into knowledge

A literature review is more than collecting papers: it is assembling a jigsaw puzzle. In analysis, break down each study into individual pieces of information. In synthesis, rearrange those pieces until patterns emerge, assumptions are revealed and new connections appear. Effective synthesis turns data into understanding, and insightful reviews can themselves advance knowledge. Major scientific contributions often arise from fresh interpretations of existing work.

Collect supporting and contradicting data ... and test theories

Carefully gathered and curated datasets enable us to evaluate existing models, identify knowledge gaps and generate new hypotheses. Comparing conflicting results reveals underlying assumptions and limitations, laying the groundwork for new experimental work. These datasets also serve as valuable benchmarks for validating observations during the experimental phase.

Inform experimental design

Experiments require prior knowledge: you must know what to measure to design an effective setup. A thorough literature review clarifies key variables, expected ranges and appropriate measurement scales (time, length, temperature, stress, etc.), resolving the apparent circularity of experimental design.[8]

Avoid redundant efforts

In the absence of a thorough literature review, we risk reinventing the wheel. Indeed, *those who ignore the past are condemned to repeat it.*[9]

Moving forward

Comprehensive, critical reviews save time, guide the allocation of resources and ensure that new research builds meaningfully on what is already known. Furthermore, a systematic literature review, together with the careful curation of diverse datasets into a coherent framework, lays the foundation for transformative ideas and fertile ground where scientific breakthroughs take root.

Exercises

1 Identify ~10 of the most important references related to your current research topic, including foundational classics and relevant works from related fields.
2 Create a comprehensive and dynamic literature review that not only informs your project but also contributes to advancing knowledge in your field. Consider the following steps:
   - Start with classics to understand fundamentals, then explore recent developments and related disciplines.
   - Summarize clearly using tables, charts or diagrams.
   - Collect and analyze datasets to identify trends, gaps and ranges for experiments.
   - Test theories and frame hypotheses by comparing findings and noting contradictions.
   - Organize the information systematically – like assembling a puzzle. Consider creating a knowledge tree or mind map.
   - Write a critical literature review: clear, concise, logically organized and unbiased.
   - Cite as you go to avoid misattribution and plagiarism.
   - Update continuously as new work emerges or feedback is received.

Notes

1 The first word in the Quran is ‘Read’ to encourage learning.
2 W. Shakespeare (1564–1616), English playwright and poet; the line is spoken by Antonio in *The Tempest* (ca. 1610).
3 See essay on *Research Teams: Creative Synergy (Section 2).*

4 This is an example of Pareto's principle: Most of the consequences are produced by a small fraction of variables involved in the process; in V. Pareto's *Cours d'économie politique* (1896).
5 The picture shows R. Feynman's last blackboard (February 15, 1988). Caltech Archives.
6 See essay on *A Great Question: Where Good Science Begins (Section 1).*
7 I. Newton (1643–1727) in a letter he wrote to R Hooke in 1675.
8 See essay on *Measurement and the Measurand (Section 3).*
9 Paraphrasing J. Santayana (1863–1952), Spanish-American philosopher, from *The Life of Reason* (1905): "Those who cannot remember the past are condemned to repeat it".

Essay 5

ELEGANCE IN EXPERIMENTAL RESEARCH

Simplicity is the ultimate sophistication[1]

Experimental researchers fall between two extremes. "Operational experimentalists" use instruments without engaging with their underlying foundation. "Principled experimentalists" deeply understand the tools, the physics involved, the imposed conditions and the assumptions behind data analysis. Striving for full engagement in every aspect of an experiment lays the foundation for robust and potentially more impactful research.

Observing excellence

A brief visit to a laboratory can often reveal meaningful insights about the team's style and spirit. The physical space, the atmosphere and the behavior of its members clearly indicate the group's approach to experimental work. Consider three dimensions:

- *Lab culture.* Outstanding laboratories cultivate a strong ethos. They uphold academic integrity, emphasize reproducibility through detailed testing protocols and foster mentorship and collaboration so that collective knowledge and practices are passed on and refined over time.
- *Well-organized and maintained facility.* Such qualities reflect a disciplined and attentive team. Conversely, disorganized and cluttered benches, piles of unused or half-dismantled devices, scattered trash and a careless attitude toward safety[2] all indicate a lack of rigor in laboratory operations. (Caveat: There are credible experimentalists who operate in seemingly messy labs…).

DOI: 10.4324/9781003727521-6

- *Elegant experiments.* The design of setups reveals the level of experimental sophistication: clarity of design, meticulous execution and innovative spirit. Some setups may work, but if they are not elegant, it is often better to start fresh with a bold new design![3]

Elegant experiments

The old saying *beauty is in the eye of the beholder*[4] reminds us that perceptions of elegance vary across individuals, cultures and time. Today, a common appreciation of style, whether in clothing, architecture or design, centers on harmony, simplicity and attention to detail. Similarly, elegant experiments share several common characteristics:

- *Focused*: Target the essence of the process under study while minimizing secondary effects.
- *Innovative, meticulous and safe design*: Introduce novel approaches or technologies to achieve research goals while ensuring operational safety.
- *Economical solutions*: Avoid unnecessary complexity, even when precise instruments are required. In many cases, less is more,[5] including low-cost operation, maintenance and repair.
- *Balanced setups*: Achieve harmonious integration and seamless operation of all components while making optimal use of materials, space and resources.
- *Well-controlled initial state and boundary conditions*: Also ensure that disassembly allows for information-rich post-test assessment.
- *Noise control*: Mitigate all forms of noise to achieve high signal-to-noise ratios.[6]
- *Careful sensor selection*: Ensure accuracy, resolution compatibility and impedance matching with peripheral systems.
- *Strategic sensor installation*: Minimize measurement interference and maximize knowledge extraction.
- *Efficient execution*: Facilitate calibration, assembly and disassembly and easy cleaning.
- *Systematic documentation*: Capture all relevant experimental details to ensure reproducibility.
- *Streamlined post-processing*: Support effective data analysis and interpretation.

Closing reflection

Elegance is difficult to define, yet easy to recognize, and it is never accidental. It arises from purpose, thoughtful design and careful attention to detail. In return, outstanding laboratories yield reliable, information-rich and repeatable data. Ultimately, elegance is not merely an aesthetic choice; it is the hallmark of great experimentalists.

Exercises

1 Visit laboratories and ask researchers to explain the experiments they are conducting. Observe the setup, process and methodology. Evaluate the experiment: Is it elegant? How could it be improved? Remember that less is often more; identify and remove unnecessary complexity. Note detractive features such as complicated or unreliable procedures, inefficient use of materials, unbalanced designs, poor boundary control, excessive noise, inadequate sensors or installation and safety issues.
2 Examine your own laboratory with fresh eyes. Is it an elegant space? Consider organization, maintenance, experimental setups and lab culture.

Notes

1 L. da Vinci (1452–1519), Italian painter and polymath.
2 See essay on *Safe Laboratory Operation: Discipline Sustains Discovery (Section 4)*.
3 N. Etchegoyen (1948–2010), Argentinian mechanician.
4 W. Shakespeare (1564–1616) wrote in *Love´s Labour's Lost* (1598): "Beauty is bought by judgment of the eye".
5 Ancient concept traced to Chilon of Sparta (6th century BC), one of the Seven Sages of ancient Greece, and then popularized by German-American architect L. M. van der Rohe (1886–1969).
6 See essay on *Noise... What a Pain (Section 4)*.

Essay 6

IN PURSUIT OF REALITY

Truth through trial

The mark of truth is its endurance under scrutiny

We enter research driven by curiosity to understand the world, reveal its patterns and predict its behavior. But scientific knowledge and engineering solutions do not emerge fully formed. They begin as fragile concepts that gradually gain strength through a long journey of questioning, testing, failure and refinement.

Paradigms shape what we see... stay alert

Every young researcher starts within an inherited framework. These paradigms provide tools and assumptions that make progress possible. They allow us to build confidently on the work of those before us. But they also shape our vision, directing what we notice and what we overlook. Confirmatory results reinforce prevailing theories; however, anomalies inevitably appear. At first, these discrepancies may be dismissed or absorbed through ad hoc modifications. But as anomalies accumulate, the framework grows increasingly complex and strained, revealing internal inconsistencies. When the old framework can no longer hold, a new paradigm emerges, more coherent and capable of revealing the world in new dimensions.[1] Electromagnetism, relativity and plate tectonics are salient examples. Engineering has also seen its own revolutionary shifts, from structural forms (beams, arches and cable-stayed bridges[2]) to the rise of transistors and nanotechnology.

DOI: 10.4324/9781003727521-7

Do not expect paradigm shifts everywhere. But remain alert to moments when a familiar way of thinking may be holding you back: frameworks help us see the world, yet they can also blind us.

The discipline of validation

How can we trust a theory in a complex world? Validation lies at the core of scientific and engineering progress. Evidence is necessary, but accumulating supportive results is never enough[3]; one more red apple does not validate that "all apples are red." Many confirming cases cannot compare to one decisive counterexample – finding a single green apple. Good research, therefore, requires deliberately searching for weaknesses; this means running tests that could *falsify* an idea, not only those likely to support it.[4] Science and engineering advance by eliminating incorrect theories.[5]

The most rigorous researchers learn to play Devil's Advocate with their own work.[6] They design tests that could reveal flaws, not just those that make results look good. Challenge your ideas with the same rigor you use on others' work.

In search of reality

A theory earns trust only if it explains what we observe, predicts what we have yet to see and allows others to reproduce its results. Without these qualities (causality, clarity, testability and reproducibility), we risk drifting toward pseudoscience, where claims are too vague to test, too flexible to fail or too insulated from empirical scrutiny.

Scientific and engineering progress depends on the courage to let reality confront our assumptions, disrupt our beliefs and reshape our understanding. Welcome anomalies, design tests that could prove you wrong and resist the temptation to fall in love with elegant ideas before they show their strength. Science advances not by protecting our theories but by exposing them to the hardest trials. The goal is not to defend what we believe but to uncover what is real.

Exercises

1 Choose an important concept in your field and examine its evolution to identify how validation efforts have shaped its development and acceptance.
2 Extend your study of scientific progress by reviewing the original sources cited here. Include additional perspectives, such as those of sociologist E. Durkheim and philosopher of science R. Giere, to explore how social and philosophical contexts shape the evolution of knowledge.

Notes

1 T. Kuhn (1962). *The Structure of Scientific Revolutions*. University of Chicago Press.
2 See essay on *Material, Form and Function (Section 4)*.
3 R. Carnap (1891–1970), German-born philosopher and leading figure of the Vienna Circle.
4 K. Popper (1902–1994), Austrian-British philosopher of science.
5 I. Lakatos (1922–1974) softened Kuhn's scientific revolutions and Popper's falsificationism and proposed the concept of *research programs* consisting of a core set of unchanging fundamental principles surrounded by auxiliary hypotheses that experience continuous refinement.
6 The Catholic Church established the *Advocatus Diaboli* in 1587. In contrast to God's Advocate, the Devil's Advocate is tasked to uncover flaws and biases in potential candidates for sainthood.

SECTION 2

Cognitive factors in research

Essay 7
LIFE CHOICES AND CHANCES

For he had neither benefice nor land.
He lived to study, teach, and understand[1]

The history of science reminds us that transformative discoveries are often made by those who venture into the unknown: those who open new paths rather than follow established routes. We must ask ourselves: are we contributing to new directions in thought and innovation? Such progress rarely occurs by chance. It grows from a deep understanding of foundational principles, fueled by sustained curiosity, disciplined effort and the determination to begin each day with purpose – to explore, to question and to learn.

Dedication

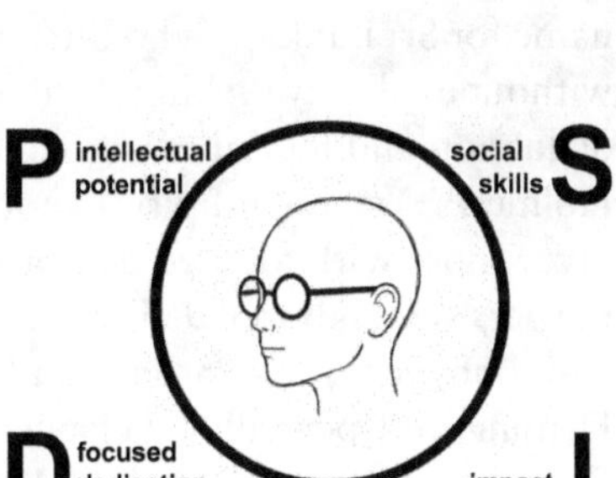

Intelligence is prized in intellectual circles, but is it really the key to meaningful impact? Consider the following exercise: think of senior individuals you know well. On a scale from 1 to 10, rate each person across four dimensions: (1) intellectual potential, (2) social skills, (3) focused dedication, and (4) overall productivity and impact. Then, analyze your data: do you see a pattern?

We asked this question to more than 100 assessors and collected data for more than 500 unnamed scientists and engineers worldwide. Then, we

DOI: 10.4324/9781003727521-9

determined the combination of IQ, social skills and dedication that determines an individual's impact.[2] Results show that dedication determines 80% of productivity and impact. Edison's quote captures it well: *Genius is 1% inspiration and 99% perspiration.*[3] It highlights that even innate talent requires hard work, through both good and bad days, to flourish. Take a moment to reflect on the insights captured in these quotes:

- Hard work beats talent when talent doesn't work hard.[4]
- Inspiration exists, but it has to find you working.[5]
- Diligence is often rewarded with discovery.[6]
- Ideas would not have come to mind had we not searched for answers with passionate devotion.[7]

Resilient perseverance

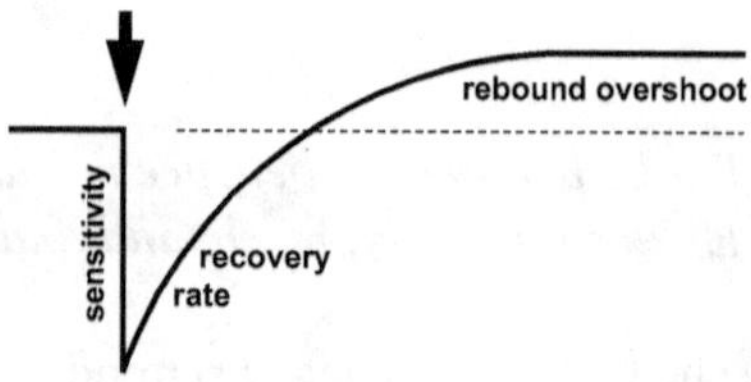

Like all aspects of life, researchers face multiple setbacks: failed experiments, data contamination, funding limitations, harsh feedback or misguided criticism. Yet some individuals persevere. They find strategies to reduce their sensitivity to negative events, to adapt, to bounce back quickly and even to overshoot with renewed commitment. In so doing, they return stronger. We use many words to describe this quality: mental fortitude, hardiness, toughness, tenacity, determination, willpower, patience, guts, conviction, drive, passion… and more recently, grit.[8]

Chance in research - Serendipity

The 5th-century tale of the *Three Princes of Serendip* (the ancient Persian name for Sri Lanka),[9] who often made discoveries through astute observation without explicitly seeking them, remains relevant. The power of observation, deduction and logical reasoning was later embodied in the figure of Sherlock Holmes.[10] Yet serendipity is not mere luck. History shows that it tends to favor those with focused dedication. Behind every "lucky" breakthrough lies the persistent labor of dedicated, meticulous researchers. Think of Archimedes and buoyancy, L. Pasteur and vaccines, M. Faraday and induction, A. Fleming and penicillin, I. Pavlov and the conditioned reflex in dogs, W.C. Röntgen and X-rays and A. Kekulé and the benzene ring. The following reflections, drawn from individuals across diverse domains, capture this idea:

- Luck is what happens when preparation meets opportunity.[11]
- Chance favors the prepared mind.[12]
- Luck has nothing to do with success. Excellence is a habit.[13]

- Chance smiles not on those who want it, but rather on those who deserve it.[14]
- In science, as in the lottery, luck favors <those> who wager the most.[15]

The environment

C. Darwin's (1809–1882) theory of evolution was shaped by his observations in rich, tropical environments, where he emphasized competitive survival. By contrast, P. Kropotkin's (1842–1921) experience in harsh Siberia led him to highlight cooperation as a key evolutionary factor. Similarly, while some researchers thrive in sterile contexts, most emerge from supportive, fertile environments. In fact, data on the educational backgrounds of Nobel Laureates support the notion that "alma mater does matter". One might argue that this correlation partly reflects the character of those who choose such institutions; yet, our own experience underscores the powerful influence of institutional culture, exemplary role models and the contagious spirit among outstanding colleagues.

Satisfaction in our choices

The history of science and engineering has been shaped by researchers who embraced intellectual inquiry as their *raison d'être*, a vocation rather than a profession. For them, living was about learning and leaving lasting contributions.

Long-term commitment to intellectual pursuits is not sustained by fleeting awards or recognition but by the satisfaction that comes from curiosity, gaining new understanding and developing innovative concepts. In the end, fulfillment does not always come to the smartest, the grittiest or the most successful, but to those who discover and embrace their true calling.

Life is complex, and the path is never simple. But you do have choices: use them wisely.

Exercises

1 Investigate the sequence of events leading to well-known discoveries in your field.
2 Read about the lives of great, inspiring scientists. Are there common patterns?
3 Write the top ten statements/quotes that resonated/inspired you the most from Ramón y Cajal's book – "Advice for a young investigator".

Notes

1 G. Chaucer (1343–1400), English poet. Description of the Scholar in *The Canterbury Tales*.
2 This is an inverse problem; see the essay on *Inversion: Inferring Causes from Effects (Section 5)*.

3 T. Edison (1847–1931), American inventor credited with foundational contributions to modern electrical and communication technologies.
4 K. Durant (1988–), American professional basketball player noted for his scoring ability in the NBA.
5 P. Picasso (1881–1973), Spanish painter and sculptor regarded as one of the most influential artists of the 20th century.
6 S. Ramón y Cajal (1852–1934), Spanish neuroscientist and Nobel laureate; quotation from *Advice for a Young Investigator* (1897).
7 M. Weber (1864–1920), German sociologist; paraphrased from *Science as a vocation* (lecture at the University of Munich, 1917).
8 A. Duckworth (1970–), American psychologist; author of *Grit: The Power of Passion and Perseverance* (2016).
9 Later adopted by H. Walpole in the 18th century.
10 A. Conan Doyle (1859–1930), British writer and physician, creator of Sherlock Holmes.
11 Seneca (4BCE–65CE), Roman Stoic philosopher and statesman.
12 L. Pasteur (1822–1895), French microbiologist; pioneered pasteurization, sterilization and vaccination.
13 D. Maradona (1960–2020), Argentine footballer widely regarded as one of the greatest in the sport's history.
14 É. Duclaux (1840–1904), French microbiologist and chemist; successor to Pasteur and contributor to germ theory and fermentation research.
15 S. Ramón y Cajal (1852–1934).

Essay 8

CREATIVITY AND CREATIVE INDIVIDUALS

Everything that can be invented has been invented[1]
(Misguided quotation)

Creativity is the generation of concepts and ideas that are novel to the individual or group involved. It is sometimes sparked by *curiosity*, as exemplified by Leonardo and Galileo. In many cases, it is driven by societal *needs*: diseases, conflicts, infrastructure and climate change. Often, it is also stimulated by *emerging opportunities*, such as those enabled by advancements in microelectronics, computing or communications.

Creativity is loosely correlated with intelligence up to an IQ of ~120[2]; beyond that, it increasingly depends on attitude and intentionality. While high performance is often expected from individuals with advanced education, prevailing educational systems and cultural norms can sometimes suppress creativity in favor of preserving the status quo.[3] Consider a basic word association exercise: write the first word that comes to mind when you read: car ______, color ______, dog ______, fire ______. Most respondents tend to offer the same set of words.[4] What does this reveal about cultural imprinting? And how can we enhance education to better support creative problem-solving?

Common characteristics of creative individuals

Cognitive science research has identified a recurrent set of traits and behaviors shared by creative scientists and engineers[5]:

- Broad knowledge and the ability to connect seemingly unrelated ideas.
- Sharp awareness of the current state of the art, without being constrained by encyclopedic detail.

DOI: 10.4324/9781003727521-10

- Selective engagement with meaningful problems.
- Ability to advance several topics in parallel, but focus deeply on one when the opportunity arises.
- Focused dedication, tremendous energy and unconditional commitment.
- Openness to emerging tools and advanced methods.
- Independent thinking and a preference for forging new paths rather than following others.
- Nonconformity, balanced with a mindful effort to avoid polarizing colleagues or organizations.
- Avoidance of energy-draining personal conflicts.
- Attention directed toward constructive rather than negative pursuits.
- Recognition of the need for system support without an obsession for total control.
- Ability to communicate clearly.
- Willingness to take risks and periodically reinvent oneself.
- Resilience in the face of difficulty, with tolerance for ambiguity and stress.
- Perseverance through failure, without excuses or self-justification.
- Ability to learn from setbacks and transform them into opportunities.
- Periodic reflection on key life events.
- Self-awareness, with a focus on leveraging strengths while minimizing the impact of weaknesses.
- Regular allocation of time for deep thinking.

Underlying processes

Creative productivity tends to peak at an intermediate age.[6] This pattern is observed across disciplines, from artists to scientists and engineers. Let's explore this trend to uncover the underlying processes.

Knowledge

Meaningful creativity requires knowledge, which typically grows with age and continues expanding well into later stages of life. Here, we use "knowledge" in a broad sense: it includes both internalized understanding and information accessed through external sources.

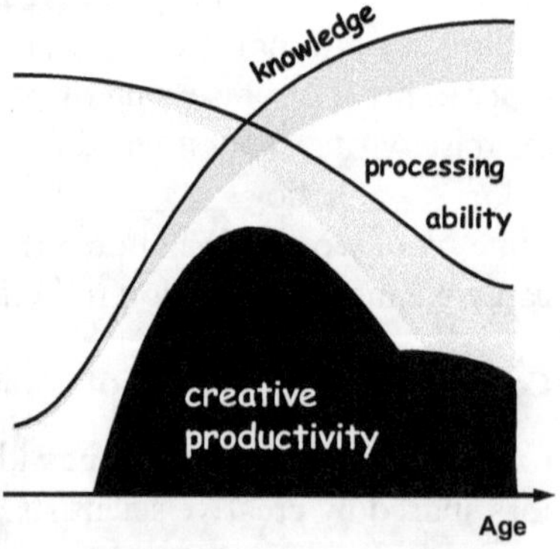

Processing ability

Creative work also relies on cognitive capabilities: unrestricted thinking, unconstrained free association, merging initially disconnected or divergent information, logical inference, organizing and visualizing information, and self-reliance on independent thought. Over time, people tend to favor depth over fluidity.[7] Problem-solving strategies become more stable, shaped by accumulated experiences and habitual thinking patterns. Meanwhile, personal preferences deepen and may gradually limit free association and divergent thinking.

Attitude and Potentializers

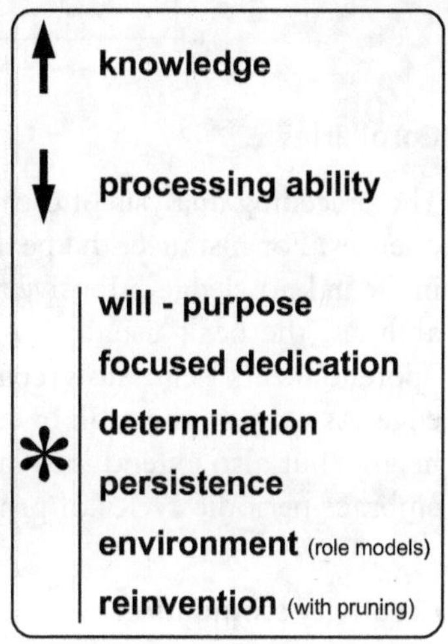

Beyond knowledge and cognitive ability, the characteristics of creative individuals reviewed above point to other factors that influence our creative productivity throughout life. Creativity requires *commitment* and *focused dedication* to the task, a *decisive attitude* and a strong *intentionality* to explore novel solutions and ideas.[8] A clear sense of *purpose* helps guide the effort toward goals, while persistent *determination* is needed to take the risks involved in making it happen.

The *environment* shapes individuals from *early childhood*[9,10] (whether through stimulating home settings, personality traits like shyness, or habits such as avid reading), continues to influence them through the *formative years* (when role models and mentors play a crucial role) and into their *current surroundings*. Some individuals thrive in competitive quasi-Darwinian environments, while others excel in austere settings where collaboration and mutual support foster growth – echoing Kropotkin's vision of cooperation.[11] Interestingly, the importance of resource availability for creative productivity may be overstated; *time* consistently emerges as the most valuable resource.[12] Ultimately, the key factor is satisfaction with the available resources.

Pruning-reinvention cycles

Finally, we observe that lifelong creative individuals go through cycles in which they trim away unproductive ideas to focus on new concepts and innovations. For example, P. Picasso (1881–1973) reinvented himself every five to seven years and remained highly productive until the end of his life.[13]

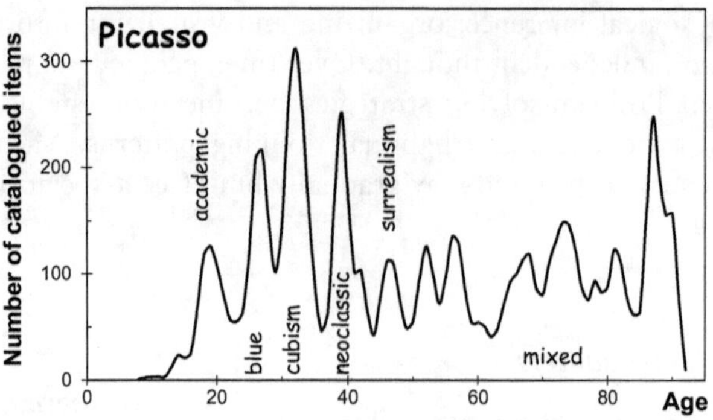

Corollaries

The preceding analysis of creativity allows us to anticipate important consequences. For instance, the peak of creative contributions tends to occur later in life in knowledge-intensive fields, while in areas requiring acute processing abilities, the peak usually comes earlier. In both cases, effectively utilizing "potentializers" can slow cognitive decline and broaden access to knowledge. As a result, the peak of creative productivity may not only reach greater heights but also extend into later stages of life – especially when individuals embrace periodic cycles of pruning and reinvention.

Creativity stimulation

Creative problem-solving often unfolds in four key stages: *preparation, incubation, illumination and verification.*[14] Among these, *illumination* remains the most elusive – the "magic step" – much like in the scientific method and engineering innovation.[15] What can we do to foster it? Meticulous analyses of logbooks kept by accomplished innovators, along with anecdotal accounts, offer valuable insights into the activities and experiences that tend to stimulate creativity. The following strategies are organized according to the core pillars of creativity discussed earlier.

Knowledge	Browse books, magazines, stores, supply shops and equipment shows. Critically read articles in your field and beyond – consider the big picture. Research your topic of interest in depth; study previous theories and discoveries. Examine anomalous evidence and identify all related facts. Explore ideas and solutions from other fields and test them. Form a network of resourceful, inquisitive individuals to tap into collective knowledge. Conduct group problem-solving exercises.

(*Continued*)

Processing	Get directly involved. Observe in detail. Ask why-questions. Immerse yourself in the problem ... "feel it". Clearly define the problem... then, transform it to reframe it entirely. Engage in critical thinking, free association and divergent thinking. Generate a wide range of ideas, hypotheses and alternatives; create mind maps. Explore the effect of various parameters; vary them across the widest possible range. Simplify, magnify, fantasize, modify and contrast; remove, substitute or add elements. Seek elegance, precision and improvement. Prototype, test, observe and describe in detail. Sketch and graph data. Keep a lab book. Systematically organize information, like solving a large puzzle. Summarize findings and identify new applications or implications.
Attitude	Select a great role model and learn from the best. Ask yourself: What would they do? Avoid conforming to accepted concepts and norms. Strive for new ideas, explore unfamiliar fields and speculate. Embrace the tension of deviating from the standard and the risks of creation. Be positive: focus on strengths. Be critical: recognize limitations. Learn from mistakes – both yours and others'. Convert negative feedback into positive action. Work hard and be persistent, driven by inner motivation and satisfaction. Have fun: seek something new every day, take on new challenges. Action leads to results... just get started!

Lifelong creativity

In most cases, creativity peaks well before any clear signs of diminished intellectual ability appear. This decline often reflects contextual pressures such as shifting family obligations, social changes or the growing responsibilities that accompany professional seniority. At other times, it stems from internal changes, including increased impatience or rigidity, diminished curiosity or motivation, conformity, an aversion to risk, fear of failure, an unchecked ego or emerging health constraints.

Aging is inevitable and affects performance along many dimensions. Yet some individuals adapt and even thrive by bringing accumulated experience and wisdom into play. Once again, lessons can be drawn from other fields. Pianist M. Argerich evolved from her early "volcanic" performances to prioritize storytelling and expression well into her eighties. Similarly, tennis legend R. Federer transformed his game in his late thirties, turning his serve into a weapon of precision and relying on optimized movement rather than raw power.

To sustain a long and creative path, remain attentive to the forces that quietly suppress innovation, recognize evolving capabilities and constraints, and remain willing to adapt and reinvent while maintaining a healthy lifestyle. Creativity is not a finite resource: the more you use it, the more you have.[16]

Exercises

1. Read this essay carefully, along with the related literature. Take time to reflect on it introspectively. Then, observe others; consider the great engineers and scientists who have shaped your field. What are the five key lessons you take away?
2. Then, revisit the essay once more – this time with a focus on identifying a few qualities you can cultivate and the strategies you can adopt to pursue a more creative path going forward.

Notes

1. A 1899 satire from the weekly *Punch* Magazine mocking the notion that technological innovation had reached its limits.
2. About 10% of the population has an IQ ≥120.
3. See: E. de Bono (1976). *Teaching Creativity*. Penguin, New York. E.S. Ferguson (1999). *Engineering and the Mind's Eye*. MIT Press, Cambridge. K. Robinson and L. Aronica (2016). *Creative Schools*. Penguin Books, New York.
4. Common answers: drive, red, loyal, hot.
5. See: S. Ramón y Cajal (1897). *Advice for a Young Investigator*. MIT Press. R.W. Hamming (1997). *The Art of Doing Science and Engineering: Learning to Learn*. Gordon and Breach. H. Gardner (1997). *Extraordinary Minds*. Basic Books, New York.
6. Best observed when individuals' longitudinal data are stacked together to observe representative trends for a group within a specific field.
7. M. Csikszentmihalyi (1997). *Creativity: Flow and the Psychology of Discovery and Invention*. Harper Perennial, New York.
8. T.M. Amabile (1996). *Creativity in Context: Update to 'The Social Psychology of Creativity'*. Westview Press.
9. D.K. Simonton (1988). "Creativity, Leadership and Chance" in *The Nature of Creativity*, ed. R.J. Sternberg, Cambridge University Press, New York.
10. K.H. Kim (2016). *The Creativity Challenge: How We Can Recapture American Innovation*. Prometheus Books, New York.
11. Refer to the essay on *Life Choices and Chances (Section 2)*.
12. B. Reitan (1996). *Creativity and Innovation in Research Groups*. Swedish Council for Building, Stockholm.
13. Similar observations are extracted from "longitudinal analyses" of individuals across many disciplines.
14. G. Wallas (1926). *The Art of Thought*. Hancourt, Brace, NY.
15. See essay on *Hidden Magic in Science and Engineering (Section 2)*.
16. M. Angelou (1928–2014) American poet. Original quote: *You can't use up creativity. The more you use, the more you have.*

Essay 9

HIDDEN MAGIC IN SCIENCE AND ENGINEERING

Knowing is not enough; we must apply it[1]

Progress in science and engineering is rarely linear. It winds through false starts, blind alleys and unexpected returns to the beginning. Reasoning is often clouded by hidden assumptions or cognitive shortcuts. Two disciplined ways of thinking, the scientific method and the engineering method, help us to navigate this crooked path. They differ in purpose but share a surprisingly common thread. Let's explore...

Two disciplines... one logic

The scientific method traces a structured path to knowledge through observation, experimentation and reasoning. Though rooted in Greek philosophy, it crystallized during the Scientific Revolution (16th to 18th centuries) and was refined by thinkers such as G. Galilei, F. Bacon and R. Descartes. While often portrayed as a tidy sequence of steps, the process is, in practice, both dynamic and iterative. Its core stages are[2]:

1 *Observe* the phenomenon carefully and objectively.
2 *Propose testable hypotheses* or plausible explanations.
3 Design and conduct *experiments* to test these hypotheses, isolating variables, minimizing biases and including controls.
4 *Analyze results* and identify which hypotheses survive empirical scrutiny.
5 Extract *conclusions* that advance understanding.

DOI: 10.4324/9781003727521-11

Engineering problem solving seeks to transform ideas into reliable, practical solutions within real-world constraints. It balances performance, safety, cost and environmental responsibility. The process also unfolds in five broad steps[3]:

1. *Define the problem*, requirements and constraints; gather essential information.
2. *Propose potential engineering solutions*; sketch, combine, adapt and resolve conflicts.
3. *Test and evaluate* these solutions under simulated or field-like conditions.
4. *Optimize* the design to meet performance targets.
5. *Monitor* real-world performance and refine through maintenance or retrofit.

The magical step

Both methods share a deep commonality. The generation of hypotheses or engineering solutions, often treated as a routine procedural step, actually depends on imagination and creative insight. It arises from deep engagement with the problem, interwoven with periods of brainstorming and mind-wandering that give rise to new ideas. Indeed, both approaches closely mirror the stages that cognitive psychologists have identified in creative problem solving.[4]

Both scientists and engineers benefit from developing multiple working hypotheses or alternative designs.[5] This broadens perspective, reduces biases and prevents premature convergence on a single path. Creativity is not chaos but disciplined openness to possibility.

Crooked but captivating

Methods provide structure, but true progress requires the courage to question, revise and adapt. Discovery flourishes through the interplay of imaginative minds that push boundaries and skeptical colleagues who probe for weakness and demand rigor. Behind every "Eureka!" moment lies a trail of wrong turns and the quiet persistence that drives science and engineering forward.

Exercises

1. Write a short letter to a curious teenager explaining how the scientific method or the engineering problem-solving process works. Make it vivid and relatable.

2 Identify and describe the key strategies you use, or could develop, to generate new scientific hypotheses or innovative engineering solutions. How do you stimulate your creativity?

Notes

1 J.W. von Goethe (1749–1832), Most influential German writer.
2 T. Winograd and F. Flores (1986). *Understanding Computers and Cognition*. Addison-Wesley Publishing.
3 E. Torroja (1958). *Philosophy of Structures*. University of California Press (English translation of *Razón y Ser de los Tipos Estructurales*, 1957).
4 See essay on *Creativity and Creative Individuals (Section 2)*.
5 Proposed by T.C. Chamberlin (1843–1928), American geologist and educator, best known for his pioneering studies on glacial geology.

Essay 10

EXPERIMENTS AND OUR DYNAMIC BRAIN

Insight grows through repeated, thoughtful inquiry and careful observation[1]

Our bodies adapt with remarkable efficiency. Muscles and bones strengthen in response to repeated exercise, especially in areas subjected to higher cyclic strains.[2] With time and consistent stimulation, they become more energy-efficient, robust and capable.

Our brains are not different. Every experience leaves a trace, particularly as we become more intimately involved.[3] Over the past century, experimental research has uncovered the processes underlying the plasticity of the brain's neural network: how it reshapes itself through learning. Some changes occur at the molecular level, such as neurogenesis, increased dendritic density and the formation of new synapses. Others affect overall brain structure and may trigger functional reorganization.[4]

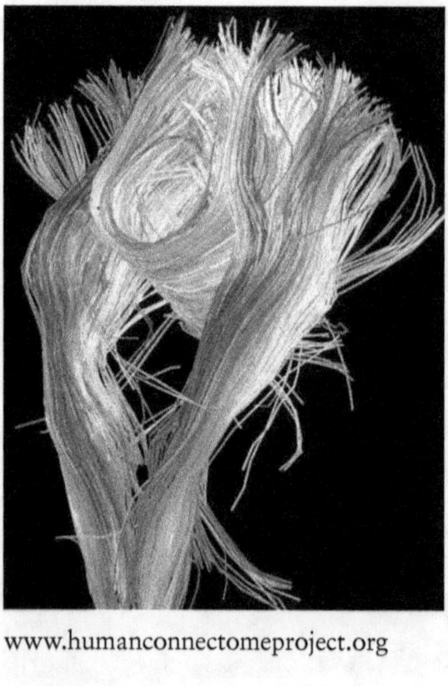

www.humanconnectomeproject.org

The brain also optimizes its performance when exposed to repeated stimuli. Selected pathways are strengthened through synaptic reinforcement and myelination, enhancing signal transmission and information processing. Synaptic connections that remain unused are

DOI: 10.4324/9781003727521-12

pruned over time. These neurological processes explain the link between practice and skill development.[5]

The outcome is truly impressive! How else could we play squash if we had to constantly reanalyze the position of our opponent, the way the racket impacts the ball and the ball's trajectory? What about driving a car or navigating the complexities of human interactions? In fact, today's AI revolution is a testament to the power of algorithms inspired by neural processes.

From chess to engineering and science, these observations underscore the importance of a rich environment and exposure to a large number of "games" and diverse experiences in developing expertise: many short "games" or cases help test ideas and recognize patterns; however, long "games" or studies complemented by thorough analysis are essential for long-term improvement.

Eventually, neural adaptation from extensive exposure enables us to internalize complex physical realities in our own brains… and then we can even conduct experiments in our minds! In fact, the most valuable byproduct of physical experimentation is the neural assimilation of the external reality being studied. It is this internalization that drives the development of expertise.

Exercises

1 Implicit learning: Our brain forms neural pathways through repetition, allowing us to automate tasks. Try brushing your teeth while focusing on each step (use your opposite hand for an added challenge). What do you notice?
2 Mind experiment: Visualize a tennis ball dropping from a height of two meters onto a ceramic floor and bouncing several times; mimic the expected sound with your mouth and record it. Then repeat the process, imagining the ball bouncing on a carpeted floor. Next, conduct the actual experiment a few times and repeat the mental version. Compare the timing and sensory details between the real and imagined experiments. Summarize your observations and insights.

Notes

1 Rooted in Aristotelian philosophy. The frequently quoted "We are what we repeatedly do" is not Aristotle's wording, but W. Durant's elegant summary of his thought.
2 See: K. Piekarski and M. Munro (1977). Transport mechanism operating between blood supply and osteocytes in long bones. *Nature*, 269, 80–82. G. Goldsztein and J.C. Santamarina (2004). Solute transport during cyclic flow in saturated porous media. *Applied Physics Letters*, 85, 2432–2434.
3 Confucius (551–479 BC): "I hear and I forget. I see and I remember. I do and I understand".

4 Salient studies include the work of S. Ramón y Cajal (1852–1934; Nobel Prize 1906), C. Sherrington (1857–1952; synapsis), E. Kandel (1929–; Nobel Prize 2000, synaptic changes during learning), R. Sperry (1913–1994; Nobel Prize 1981, lateralization of brain function).

5 D. Eagleman (2020). *Livewired: The Inside Story of the Ever-Changing Brain.* Pantheon.

Essay 11

TOOLS FOR REASONING

Disciplined thought

The things you think about...
determine the quality of your mind[1]

Scores on intelligence tests have steadily risen over the last century. The average person scoring ~100 today would score around 110–115 on a test from a hundred years ago, placing them in the 74th to 84th percentile.[2] This phenomenon is known as the Flynn effect. We observe a similar pattern in other domains, particularly in Olympic and professional sports: world records fall, performances surge and athletes routinely redefine what once seemed physically impossible. Clearly, these gains cannot be explained by Darwinian evolution: the human genome has not changed enough in just two or three generations. So what has? The answer lies elsewhere: enhanced education, optimized training, carefully designed routines, better diets and nutrition, detailed performance monitoring and more stimulating environments.

Cultivating the mind

The knowledge-creation strategies developed over the last 3,000 years have gradually permeated both our culture and educational systems.[3] Today, educating future researchers goes beyond transferring information: it also cultivates *tools for reasoning*.[4] How are we taught to think? Let's explore strategies we can actively apply in daily routines and educational programs to train the mind for insight, rigor and creativity:

DOI: 10.4324/9781003727521-13

- *Logical thinking*: Identify causal relationships, analyze problems and synthesize findings. Remain skeptical, test theories by reducing arguments to absurdity, identify counterexamples and falsify claims.[5]
- *Scientific method*: List all relevant facts related to a topic of interest. Conceive multiple hypotheses that can explain observations, then contrast these hypotheses against the data. Let the hypotheses guide further fact-finding and predict events that have not yet been observed.
- *Words – Storytelling*: Construct a concise, compelling and physically sound story that captures the object of study, whether describing a physical system, a historical process or a social interaction – like assembling a dynamic puzzle whose image is not predefined but gradually reveals itself as the essential aspects become apparent.[6]
- *Metaphors and analogies*: Identify inspiring stories, vignettes and persuasive examples across disciplines. For example, relate water polarity to microwave heating, osmosis-driven flow to the gastrointestinal tract's response to seawater, carrots to pile foundations, and breathing to atmosphere–ground interaction.
- *Fundamental physical concepts*: Understand physical processes at their inherent scale. For instance, relate heat to molecular vibration,[7] vibrating molecules to ionic diffusion… and explain irreversible thermodynamics!
- *Thought experiments and "what if" scenarios*: Play a sequence of causally linked events in your "mental simulator"; explore possible outcomes and constraints.
- *Physical empathy*: Embody the system you are studying. Imagine yourself as a molecule to understand hydrates, as a bacterium during colonization, as a moving charge within an electric current, as a shell when designing structures or as a grain to grasp granular flow in an hourglass.
- *Case-based learning*: Remember the past to avoid repeating mistakes. Analyze case histories to gain new insights that improve theories and solutions.
- *Physical sketches – Attention to detail*: Prepare physically correct sketches, interaction diagrams and flow charts that capture the essence of the phenomenon under study.[8] This exercise will focus your mind and encourage detailed observations.
- *Graphical solutions*: When possible, attempt graphical rather than purely numerical or analytical solutions (e.g., Laplacian fields). Graphical approaches highlight boundary conditions and draw attention to the system's most critical features.
- *Experimental studies*: Run experiments every day. Each experiment captures the full physical reality, in contrast to analytical or numerical abstractions. Ensure experiments satisfy similarity and boundary conditions. Identify measurement biases and limitations.

- *Analytical solutions*: As an experimentalist, do not shy away from expressing physical observations in mathematical terms. Analysis can guide experimental design and illuminate the interplay among governing variables. But proceed with care and document and list all assumptions along the way.
- *Numerical solutions*: Become comfortable with numerical methods so that you can explore complex interactions and boundary conditions and upscale or extend the scope of your experiments.
- *From databases to knowledge – AI tools*: Collect data for your field of interest, leverage automation and incorporate pre-existing databases. Then, harness statistical and AI tools to mine new information and insights from the data.

Synergy among tools

There is a profound synergy among these "tools for reasoning". Their value does not rely on sophistication; even elementary, disciplined practices can gradually cultivate deep intuition over time. In an era where powerful computational tools may convey a false sense of precision, maintaining a cautious and critical mindset is essential. Order-of-magnitude estimates, scaling analysis and physically grounded reasoning remain indispensable foundations, guiding us through the complexity of modern science and engineering.

Exercises

1 Identify other "tools for reasoning" that can be incorporated into training routines.
2 Prepare a syllabus for a course you would like to teach in your field, embedding "tools for reasoning". Use a two-column format: in the left column, list the technical content organized into units; in the right column, specify the most appropriate reasoning tools you would embed within each unit.

Notes

1 M. Aurelius (121–180), Roman emperor and Stoic philosopher.
2 Note: IQ tests are normalized to render a mean μ_{IQ}=100 and standard deviation σ_{IQ}=15.
3 See essay on *Knowledge Creation Throughout History (Section 1).*
4 D. Dennett (1942–2024), American philosopher and cognitive scientist – *Tools to Transform our Thinking.*
5 See essay on *In Pursuit of Reality: Truth Through Trial (Section 1).*
6 See essay on *Writing: Crafting Meaning One Piece at a Time (Section 6).*
7 Molecular vibration takes place at ~10^{13} Hz with amplitudes in the order of 10^{-3}nm.
8 See essay on *Sketches and Plots: Visual Thinking and Communication (Section 6).*

Essay 12

BIASES IN DECISION-MAKING

It is a capital mistake to theorize before you have all the evidence. It biases the judgement[1]

Despite our confidence in being rational thinkers, our minds are prone to pervasive fallacies in perception, information processing and decision-making. These inherent cognitive limitations have far-reaching implications, affecting not only our everyday judgments but also the effectiveness and objectivity of our research. Recognizing and understanding these constraints is essential for improving the quality of both critical thinking and problem-solving.

Preconception and perception

Perception cannot occur without preconceptions, yet these very preconceptions bias what we perceive.[2] In fact, our minds rely on existing mental frameworks to fill in gaps in perception. As a result, we face the ongoing challenge of discerning what is real beyond our preconceptions and beliefs.

In building those mental frameworks, our brains tend to favor new information that aligns with what we already know or believe, while downplaying or dismissing contradictory data, a tendency known as *confirmation bias*. As supporting evidence accumulates, resistance to change increases, and dissenting information becomes less impactful. Ultimately, s*cience progresses when observations compel us to revise our preconceptions.*[3] Although all fields are influenced by preconception biases, some areas – such as the study of gender and racial differences – remain especially vulnerable.

DOI: 10.4324/9781003727521-14

Biased processing of feedback

Decision-making is shaped by the outcomes of prior decisions and past experiences. Related biases include relying solely on outcomes (i.e., using partial or incomplete information),[4] better recollection of the earliest or latest data in a sequence (primacy and recency effects),[5] success/failure attributions (i.e., the tendency to attribute success to one's own skills and blame failures on chance), logical fallacies in recall (i.e., the tendency to feel confident in a "logical" reconstruction of an event) and hindsight bias (i.e., the lack of surprise after an event has occurred). Biased processing of feedback from past experiences hinders the development of true expertise.[6]

Limited capacity

Cognitive constraints also shape decision-making. According to Miller's Law,[7] our working memory can hold only 7±2 items. As a result, we often base decisions on just a few cues: typically, only three cues account for 80% of the variance in an individual's response.[8] Once a choice is made, we tend to selectively seek additional data that confirms our position, rather than challenging it. This creates a false sense of certainty. Consequently, more data may provide more confidence, but it does not necessarily improve the quality of our decision.

Poor intuitive statisticians

Biases and limitations also affect decisions involving randomness and frequency, from hurricanes and traffic to clinical trials:

- *Gambler's fallacy and misperceptions of chance fluctuations*: We tend to believe that past events in a random sequence will influence future outcomes; for example, if the roulette wheel landed on red in the last four spins, we "feel" the next one must be black. Conversely, we may feel that a frequently occurring event not recently observed is more likely to happen next.
- *Inability to conceive randomness*: We struggle to create truly random sequences and often look for patterns where none exist, a cognitive bias known as apophenia.[9]
- *Poor estimation of events with extreme probabilities*: We misjudge the likelihood of very rare or very common events.
- *Conjunctive and disjunctive events*: We overestimate the probability of events that must occur together (conjunctive: $P(A \text{ and } B) = P(A) \times P(B)$ if A and B are independent) and underestimate the probability of one of

several events happening (disjunctive: $P(A \text{ or } B) = P(A) + P(B)$ if A and B are mutually exclusive).[10]

- *Illusory correlation*: We falsely perceive relationships between events that frequently occur together.
- *Inability to recognize regression toward the mean*: Extreme results tend to return to the average, contrary to people's intuition; for example, children of two tall parents are likely to be shorter than their parents.
- *Sample size – Law of small numbers*: We often overlook how sample size affects the variance of the mean, and we mistakenly believe that small samples accurately represent larger populations.[10]

Misuse of heuristics

Decision-making can be negatively affected by the misuse of "heuristics" or mental shortcuts we have developed to expedite problem solving[9]:

- *Representativeness*: The probability of event A being related to event B is judged on how much A resembles B. This leads to the unwarranted conclusion that a good fit implies causality.
- *Availability*: Instances with a large number of cases are more easily and quickly recalled than those from less frequent categories, potentially skewing judgment.
- *Insensitivity to assumptions*: Expedited decision-making may disregard assumptions made in earlier analyses or fail to recognize unstated assumptions. For example, design changes during construction often increase risk, as seen in the 1981 walkway failure at the Kansas City Hyatt Regency Hotel.
- *Adjustment and anchoring*: As discussed above in the context of preconceptions, individuals form their beliefs based on an initial reference point and adjust them as new information becomes available.[9] However, adjustments tend to be conservative or insufficient, in part because we are committed to earlier ideas. Anchoring affects all decision-makers, including both engineers and scientists. It is, therefore, not surprising that paradigm shifts are often introduced by researchers new to a field.

Anchoring has other manifestations in engineering and science. We often place significant trust in prior findings by respected figures, sometimes even doubting our own observations or disregarding dissenting data when they conflict with established trends. This reliance on past results or "anchors" can quietly hinder scientific progress. The measurement of the speed of light provides a notable example: for decades, published values clustered around earlier incorrect results, and the reported uncertainty ranges often excluded the value we now know to be correct. Another prominent case is the measurement of the charge of the electron: R.A. Millikan's oil droplet experiment, which earned him a Nobel Prize, had a strong anchoring effect. More accurate results gathered later were frequently dismissed as flawed, while values that closely matched Millikan's original findings remained favored.[11] These examples remind us that scientific advancement requires not only rigor but also the courage to question even our most trusted assumptions.

Exercises

1 Review the essay thoroughly and explore additional literature on the topic. Reflect on your own thinking to become aware of potential cognitive biases that may influence your judgment.
2 Identify the five cognitive biases that most affect you personally. For each bias, provide examples from both your everyday life and your experiences in engineering or scientific work.
3 Humans are notoriously poor intuitive statisticians. Can you propose an evolutionary explanation for this tendency?
4 We often accept information as personally relevant, even when it could apply to most people (Barnum effect). How do you react when you read fortune cookie messages or hear the results of casual personality tests?

Notes

1 S. Holmes in *A Study in Scarlet* (1887) by A.C. Doyle (1859–1930), Scottish author.
2 A. Nin (1903–1977) French-American writer: "We do not see things as they are. We see them as we are".
3 Paraphrasing V. Rubin (1928–2016), American astronomer.
4 See essay on *Statistical Concepts for Experimental Design and Execution (Section 3).*
5 I.L. Janis and L. Mann (1977). *Decision Making, A Psychological Analysis of Conflict, Choice and Commitment.* The Free Press, New York.
6 R.M. Hogarth (1980). *Judgement and Choice, the Psychology of Decision.* John Wiley and Sons, Inc., New York.
7 G.A. Miller (1956). The magical number seven, plus or minus two: Some limits on our capacity for processing information. *Psychological Review*, 63(2), 81–97.
8 P. Slovic and S. Lichtenstein (1971). Comparison of Bayesian and regression approaches to the study of information processing in judgement. *Organizational Behavior and Human Performance*, 6, 649–744.
9 R.M. Hogarth (1975). Cognitive processes and assessment of subjective probability distributions. *American Statistical Association*, 70, 271–289.
10 A. Tversky and D. Kahneman (1974). Judgment under uncertainty: Heuristics and biases. *Science*, 185, 1124–1131.
11 R. Feynman (1985). *Surely You're Joking, Mr. Feynman!*, W.W. Norton & Company.

Essay 13

BIASES IN COMMUNICATION

We hear not what is said, but what our minds are prepared to believe[1]

Communication goes far beyond the transmission of information. It is a complex process in which a message is encoded by the sender, transmitted and then decoded by the receiver. Crucially, both sender and receiver bring their own cognitive filters, cultural assumptions and emotional contexts into the exchange. This leads us to reflect on the gap between what was said, what was heard and what was truly understood!

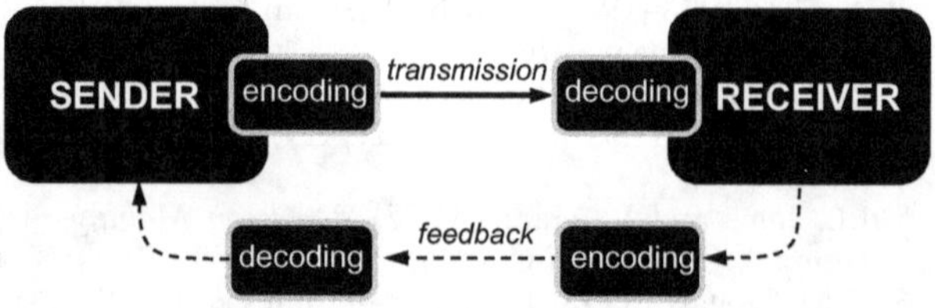

True understanding requires more than linguistic accuracy; it demands shared interest, continuous feedback and the gradual alignment of attitudes and knowledge. While education helps establish a common ground, challenges such as information overload, distractions, noise and internal or environmental stressors persist.

The words we use...

Language does more than convey information: it shapes how we perceive, classify and judge the world. It evolves continuously in both everyday conversation and technical discourse, and we tend to underestimate the extent to which word choice influences thought. Consider, for example, the growing

DOI: 10.4324/9781003727521-15

recognition of the impact of "gendered languages" on biases and stereotypes and the shift toward gender-neutral forms (e.g., fireman → firefighter).

The power of language lies not only in what it describes but also in what it implies, for example, through the emotional and ideological framing of terms. *Clean coal* is a rhetorical term, not a scientific reality. A *puppy* elicits tenderness; a *juvenile dog* does not. In debates over reproductive rights, the same biological reality becomes a *fetus* in one context and an *unborn baby* in another, each term carrying its own moral weight. Political and social conflicts magnify this effect: dehumanizing labels such as *savage, alien* or *illegal* are deployed to diminish and distance others, reshaping public perception long before facts enter the conversation.

Cultural norms add another layer of complexity. Politeness conventions differ widely: for some, "not bad" means very good, an "interesting idea" actually indicates a poor solution and the euphemistic phrase "passing away" means dying. These examples reveal how deeply culture is embedded in language.

A different linguistic challenge arises when misnomers or labels are created before phenomena are fully understood. They can persist for centuries, shaping collective memory even when they distort reality. The "Dark Ages", for example, overlapped with the Golden Age of Islam, a period of extraordinary scientific and cultural achievement. Yet once adopted, such terms become convenient reference points, resistant to correction. Indeed, you will not get *French fries* if you order *Belgium* or even *Chilean fries* – their actual origin!

These examples highlight that language is never neutral. It frames our interpretations, encodes our assumptions and influences how we interact with one another. To be attentive to word choice is not merely to use language well; it is to recognize how deeply it shapes our understanding of the world.

Communication in engineering and science

Technical language presents similar difficulties:

- *There are many misnomers*: A jellyfish is not actually a fish; the Big Bang was not an explosion; there are no "holes" in black holes; and the "imaginary" part of a Fourier transform is, in fact, very real!
- *Many terms are gendered*: How might J. Bronowski's illuminating 1973 book *The Ascent of Man* be titled if it were published today?
- *Some words carry dual semantics*: Consolidation means pressure diffusion for a geotechnical engineer, whereas consolidated sand implies cementation for a petroleum engineer. Likewise, clay refers to submicron particles for soil mechanicians but to phyllosilicate minerals for a mineralogist.

- *Some expressions reflect intuitive but incorrect understanding*: While electrical engineers often say that current "flows through" a capacitor, in reality, almost no electrons cross the dielectric material.
- *Incorrect terms affect perception and subsequent analyses*: Fine-grained sediments that can form a mud ball are often classified as cohesive soils even though there is no "cohesive strength" in remolded mud. Not surprisingly, most infrastructure failures involve fine-grained soils!

The point is clear: precise technical communication is critical in engineering and science, and the misuse of language can have long-lasting negative consequences.

Communication = sharing (thoughts, concepts, information, emotions, understanding)

Effective communication is fundamental to achieving successful outcomes, facilitating collective knowledge and enabling informed decision-making. Misunderstandings and miscommunications, by contrast, can undermine interpersonal and professional relationships, damage work environments, lead to accidents and contribute to major historical events, including political upheavals and international conflicts.

Technical and scientific fields are not immune. Data from civil engineering case studies indicate that roughly 25% of failures can be traced to communication breakdowns.[2] Poor communication between design and inspection teams played a major role in the Teton Dam failure.[3] And NASA's Mars Climate Orbiter was lost because the two teams involved used different units – JPL used metric units while Lockheed Martin used imperial units.

Communication within organizations

The vast majority of communication in large organizations is sub-formal. Communication networks tend to become more effective as relationships among individuals grow more stable over time.

Horizontal and diagonal communication are often distorted by self-interest. Unfavorable information is frequently forgotten, and when it is communicated, it is often rejected or sent back. Individuals who bring forward disturbing information are quickly labeled as "troublemakers" and may be isolated or removed from the organization.

Vertical communication is typically slow. Furthermore, information passed upwards in a hierarchy tends to be distorted to align more closely with what senior members want to hear. Information that benefits a senior individual is often exaggerated, while unpleasant details are minimized.

Since information affects prestige and power, one can expect continuous distortions and filtering. Individuals may adopt counter-biasing strategies, such as reducing reliance on a single source and consulting multiple ones. However, these approaches increase information overload and slow down processing. Similarly, organizations may assign assistants to manage information flow, yet such measures can introduce further distortions, omissions and delays. This often aggravates the very problem they intend to solve.

Critical lesson

Ultimately, the most effective way to foster meaningful communication is through the alignment of goals and values: when individuals genuinely share a purpose, communication becomes clearer, more resilient and more trustworthy.

Exercises

1 Identify common misnomers or misleading terms in your field. Briefly explain why they are inaccurate or problematic.
2 Identify the three communication biases that most affect your interactions. Then, outline a four-to-five-step strategy you can use to improve your communication skills in light of these biases.
3 Choose a team or organization you know well. Analyze its communication dynamics and identify the most harmful communication biases at play. Reflect on their impact and consider possible ways to address them.

Notes

1 Paraphrasing ideas associated with F. Bacon (1561–1626), English philosopher and pioneer of the scientific method, and W James (1842–1910), American philosopher and psychologist. The common quotation about miscommunication, "I know you think you understand what you thought I said, but what you heard is not what I meant", has no confirmed source and is variously attributed to R. McCloskey, R. Nixon and A. Greenspan, among others.
2 J.C. Santamarina and C.L. Turkstra (1989). *Human Factors and Communication Problems in Foundation Engineering: Current Principles and Practices*, ed. F.H. Kulhawy, vol. 2, pp. 857–868.
3 J.L. Sherard (1987). Teton Dam Failure. *Engineering Geology*, 24, 283–293.

Essay 14

RESEARCH TEAMS

Creative synergy

The whole is greater than the sum of its parts[1]

Creativity sparks in the individual mind, but it never happens in a vacuum. It draws on the accumulated knowledge that is part of society's cultural wealth. This is the social dimension of creativity. We are part of an immense network striving together to build better lives. Isn't it astonishing what simple bees or ants can accomplish as a team? Now look around: your home, running water, transportation, your computer, the food on your table... even the internet and Wikipedia! There are no limits to what we can achieve when we collaborate.

Teamwork can elevate each individual's performance and produce surprising outcomes. Imagine a group of potters working around a table, each starting with a lump of clay.[2] Every few minutes, they pass their work to the person next to them to continue shaping it. The piece evolves with each exchange: early stages define the overall form, and later participants refine the details. Eventually, the emergent final pieces go beyond the imagination of the initiators, capturing the collective results of many hands and perspectives coming together. Research efforts follow a similar path: initial ideas evolve as they interact with inspiring collaborators (and later with anonymous reviewers) whose expertise helps refine arguments, challenge assumptions, and strengthen the final product.

DOI: 10.4324/9781003727521-16

Large-scale, complex, multidimensional problems demand teamwork to bring together the small-scale contributions of focused and dedicated individuals. This applies from the team of sculptors working with G. Vigeland at Frogner Park in Oslo,[3] to the team of scientists and engineers working on planetary exploration. But how do we build an effective team? How do we select the musicians and lead them to form a coherent orchestra?

Recognizing talent

Selecting the right people is the foundation of any successful team. W. Buffett emphasizes three essential qualities when choosing individuals: integrity, energy and intelligence. Other qualities matter too, such as loyalty and resilience. Who would you want beside you in a shipwreck?

The movie *Moneyball* vividly illustrates this principle through the story of a baseball team's general manager. Initially relying on conventional scouting methods, he made poor recruitment decisions. That changed when he partnered with a young economics graduate who applied statistical analysis to identify undervalued players with the specific skills the team needed, instead of investing in expensive, high-profile players the team did not really require. This data-driven approach marked the beginning of modern sports analytics and led to remarkable success. The key lesson is clear: select individuals whose abilities align directly with the team's strategic goals.

From a group to a team – leadership

A team is more than just a collection of skilled people; it is a cohesive unit working toward a common goal. As in soccer, even the brightest stars cannot win if they do not have a collaborative spirit. A foundation of trust and mutual respect is also essential. Team members must have full confidence in the integrity of each other's actions. This is especially critical in science, where any form of misconduct can undermine the team's credibility.[4]

Great leaders are inspiring motivators and catalysts for collective action. Like a good coach, leaders of research teams must recognize talent, guide individuals to maximize their strengths while minimizing their weaknesses and promote continuous learning. Team leaders plan strategically, explore multiple alternatives and remain curious and willing to take risks. The most successful leaders demand excellence, inspire their teams with humility and genuine commitment, avoid favoritism, foster collaboration across all players and transform a group of talented individuals into a cohesive, goal-driven team.

Synergism vs. antagonism

As the team evolves, members gradually find their roles.[5] Some are directly engaged in the task: contributing information, initiating ideas, seeking opinions and integrating efforts. Others embrace group-maintenance roles: encourager, harmonizer, tension reliever or gatekeeper. Then, there may be those who adopt dysfunctional roles: blocker, pessimist, aggressor, anecdote-spinner, recognition seeker, dominator, special-interest-biased member, joker or sarcastic commenter.

The balance between players and roles will determine the collective performance of the team. Synergistic teamwork stimulates all members, raising the bar together. It brings collective intelligence and the pooling of information, considers a wider range of approaches and fosters creative outcomes through cross-fertilization. Fluent communication and a group-minded attitude are required for synergistic teamwork, helping maintain the motivation needed for creative problem-solving.

On the other hand, teams with uncommitted or energy-draining antagonistic members may produce collective outcomes that do not surpass the quality of work achieved by individuals working alone. Team members might disagree on the nature of a problem, while less-capable individuals can dominate discussions through sheer persistence. Additionally, members eager for consensus may stop exploring valuable alternatives or seeking additional information, a phenomenon known as groupthink. Furthermore, teams – like individuals – tend to favor information that confirms the majority's initial position and disregard opposing evidence, demonstrating confirmation bias.[6]

Final reflection

How can you recognize an underperforming team? Pay attention to the warning signs: lack of enthusiasm, low trust, minimal encouragement, excessive criticism, an overbearing leader or a culture driven by competition and personal reward.[7] Being part of such a team can be deeply frustrating... but it can also become a turning point: these moments invite introspection and open the door to personal growth. They can push you into new, transformative roles that strengthen both your development and the team's collective performance. When the dynamic around you fails to inspire, *be the change you want to see!*[8] Set the tone. Lead by example. And above all, keep your eyes on the long game. Your commitment, integrity and openness will shape not only the outcomes you achieve but also the kind of team you help build.

Exercises

1 Research respected coaches from team sports. Study their biographies and published interviews to extract the key principles that guided their coaching approaches. Translate the lessons learned from these coaches into practical strategies for leading research teams effectively.

2 Consider your current performance within your team. What roles do you typically assume? What roles would be most valuable for you to play? Identify three concrete actions you can take to enhance your contribution and improve overall team effectiveness.

Notes

1 Aristotle (384 BC–322 BC), Greek philosopher and polymath.
2 This exercise was conducted by the British sculptor Anthony Gormley (1950–) with a team of artists in his studio.
3 G. Vigeland (1869–1943), Norwegian sculptor.
4 See essay on *Pathological Science and Misconduct (Section 7)*.
5 M. Burgoon, J.K. Heston and J. McCroskey (1980). Communication roles in small group interaction. In *INTERCOM Reading in Organizational Communication*, ed. S. Ferguson and S.D. Ferguson, Hayden, Roshelle Park, 412 pp.
6 See essay on *Biases in Decision-Making (Section 2)*.
7 See: R.L. Bailey (1978). *Disciplined Creativity for Engineers*, Ann Arbor Science Publishers, Ann Arbor, MI. D.S. Gouran and R.Y. Hirokawa (1986). Counteractive functions of communication in effective group decision-making. In *Communication and Group Decision-Making*, ed. R.Y. Hirokawa and M.S. Poole, pp. 81–92. Sage, Newbury Park, CA.
8 M. Gandhi (1869–1948), Political and social activist who employed nonviolent resistance.

SECTION 3

Experimental design concepts

Essay 15
VARIABLES

Only a few variables define a process; the rest merely explain minor deviations

Variables can be classified according to their nature and role. Some are *physical*, directly measurable properties of the natural world, such as mass, length or temperature. Others are *conceptually defined*, like intelligence or resilience, formulated in abstract terms rather than tied directly to observation; to make such concepts useful in research, they must be *procedurally/operationally defined*,[1] for instance, translating intelligence into measurable performance on an IQ test. Variables can also be classified as dimensional or dimensionless. *Dimensional* variables carry physical units, while *dimensionless* variables emerge as ratios or normalized quantities.[2]

Size dependence

An important distinction among variables is how they scale with system size. *Intensive variables* do not depend on the amount of substance: temperature, for instance, remains the same whether measured in a drop of water or in a lake. *Extensive variables*, by contrast, scale with system size: mass and volume both increase as the amount of substance increases. *Specific variables* express an extensive property relative to mass or substance, for example, molar mass or molar volume.

Size dependency has practical implications. The information in intensive variables often has broader validity and usefulness since they can be compared directly across systems of different sizes. Ratios of extensive variables

DOI: 10.4324/9781003727521-18

yield intensive variables; for instance, density arises from dividing mass by volume, and pressure is the force per unit area.

The intensive property of a mixture I_{mix} is the weighted mean of the intensive properties of the components I_x according to the extensive property m_x, assuming components do not interact. In contrast, the extensive property of a mixture E_{mix} is the weighted sum of the extensive properties of the components E_x (Euler's homogeneous function theorem):

$$I_{mix} = \frac{\int m_x I_x}{\int m_x} \qquad E_{mix} = \int m_x E_x$$

Variables in experimental studies

Variables can also be classified by their role in an experiment. Consider a study on how different fuels affect a car's speed and efficiency:

- *Independent variable*: manipulated by the researcher, for example, type of fuel.
- *Dependent variable*: measured in response to changes in the independent variable, for example, maximum speed, fuel efficiency.
- *Control variable*: held constant to isolate effects, for example, same car, same driver.
- *Confounding variable*: not controlled but still influences results, for example, differences in driver skill.
- *Extraneous variable*: random factors that may affect outcomes, for example, engine temperature at the start of each test.
- *Moderator variable*: alters the strength of the causal link between independent and dependent variables, for example, engine type (diesel or gasoline).
- *Mediator variable*: explains the causal link between independent and dependent variables, for example, combustion efficiency linking fuel type to speed.

Exercises

1 Select an experiment that interests you. Describe the experiment briefly, including its purpose and setup. Identify the main variables involved. Classify each variable according to the descriptors discussed in this essay: physical vs. conceptual, dimensional vs. dimensionless, intensive vs. extensive, and independent, dependent, control, confounding, extraneous, moderator or mediator.

2 In *The Logic of Modern Physics* (1927), P.W. Bridgman introduced the concept of operationalism: a scientific concept is meaningful only if it can be defined by the operations or procedures used to measure it. According to Bridgman, questions that cannot be connected to a measurable procedure are scientifically meaningless. Can you think of questions in your field that would be considered meaningless under operationalism? Explain why they lack measurable or observable definitions.

Notes

1 P.W. Bridgman (1882–1961), American Nobel Prize in Physics.
2 See essay on *The Dimensional and The Dimensionless (Section 3)*.

Essay 16

MEASUREMENT AND THE MEASURAND

What you want to measure…
what you think you measure…
what you actually measure![1]

Imagine a metal object and a wool scarf sitting untouched in the same room for hours. Touch them both at the same time. Which one feels colder? The metal does. But since they have both been in the same room, they must be at the same temperature. So why does the metal feel colder? Actually, we do not sense their temperature directly but rather how quickly they dissipate heat from our hand. In fact, an ice cube would melt faster on the metal plate than on the cloth, even though the metal felt colder to the touch! This distinction between what we "think we measured" and what was "really measured" captures a fundamental dilemma in experimental research.

From observation to quantification

Measuring seems straightforward at first glance: we map something observable onto a number. This is easy when the target is mass or voltage. But many concepts, such as *creativity*, *pain* or *rock quality*, are not directly measurable; in this case, measurement must begin with an operational definition that associates the measurement protocol with the measured value. Whether physical or operationally defined measurements, there are unstated assumptions the moment we assign a number to an observation: that the number is unique, that it reflects differences that matter and that it can be compared meaningfully. These are big assumptions.

DOI: 10.4324/9781003727521-19

Principles underlying every measurement

Experimentalists learn quickly that it is relatively easy to obtain "neat measurements", but getting "true measurements" can be truly challenging. The following measurement axioms provide important context:

- *Measurement is selective.* Every act of measurement chooses what to quantify. By focusing on one aspect of a system, the rest is left out. The result is an inherently partial, incomplete version of reality, curated by the experimenter.
- *Measurement affects the measurand.* It is not just observation; it is interaction. Passive measurements draw energy (e.g., thermocouples) while active measurements inject energy into the measured system (e.g., thermistors). In all cases, the act of measuring inevitably affects the measurand.[2]
- *Measurement is mediated.* All measurements are filtered through a transfer function; consequently, the recorded signal is not the original physical input. For example, a seismometer produces a voltage—time record, not the ground velocity—time history during an earthquake. Recovering the physical input requires knowledge of the instrument's transfer function.
- *Measurement is always uncertain.* There are chronic sources of errors: sensor mounting imperfections, poorly controlled or unknown boundary conditions, undetected inhomogeneities, confounders (hidden variables), inconsistent calibration and noise. These are difficult to quantify, hard to isolate and often ignored.
- *Measurement is scale-sensitive.* Experiments may operate in the wrong variable domain. If scaling laws are violated or similarity requirements unmet, then the results – though accurate within their domain – fail to reflect the behavior they aim to represent.[3] Think of a scaled model ship that looks identical to the real one: its response on water may appear insightful, yet it may not reproduce the governing physics of the full-scale prototype.

We must then conclude that *we are rarely measuring what we think we are measuring.*

What can you do?!

Clearly, all measurements are inherently uncertain,[4] and many are quietly misleading. Precision is no guarantee of truth, and confidence in the number can mask confusion about the observation being measured. Therefore, be rigorous. Question your instruments and procedures. Question your assumptions. And above all, remain agnostic of your own results.[5] The danger rarely lies in what the instrument reports but in what we believe those results mean.

Exercises

1 Tests are continuously conducted on the International Space Station (ISS) to explore phenomena under microgravity. But is it truly microgravity? After all, the ISS orbits at approximately 410 km above the Earth's surface, where gravity is still about 90% as strong as on the ground. So what is the actual gravitational environment aboard the ISS? Discuss the nature of gravity in orbit and the implications this has for interpreting scientific experiments conducted under so-called "microgravity" conditions.
2 Discuss the potential biases a psychologist may exert on an individual during a support session. Consider factors such as language, nonverbal cues, cultural assumptions and emotional reactions.
3 Identify possible confounding variables in the observed correlation between steel production and car manufacturing. Analyze whether external factors might independently influence both variables, creating a spurious or misleading correlation.

Notes

1 Source unknown; similar ideas appear in the writings of Lord Kelvin (1824–1907), N. Bohr (1885–1962) and W. Heisenberg (1901–1976).
2 This is the foundation of Heisenberg's uncertainty principle in quantum mechanics.
3 See essays on *Similarity and Scaling Laws (Section 3)*.
4 This is not a falsification of causality! – Refer to the essay on *Free to Explore Causality?* (*Section* 7).
5 A frequent cited quote alludes to this idea: "nobody believes models but the modeler... everybody believes experiments but the experimentalist".

Essay 17

EXPERIMENTAL METHODS

Diverse approaches

Design your experiments so that nature cannot mislead you[1]

Experimentation takes many forms, and its expressions vary across disciplines. Consider the caricatured views of engineering and science. *Engineers'* "how" questions drive experiments that identify critical parameters, optimize performance, validate designs and bridge theory and practice, as reflected in standards and codes. The classical trial-and-error approach, rooted in Roman times and later exemplified by T. Edison,[2] remains a common strategy in engineering, especially for rapid prototyping and system refinement. In contrast, *scientists'* "why" questions seek to reveal the fundamental principles and mechanisms underlying observed phenomena. Ideally, the scientific method involves a rigorous process of hypothesis testing, grounded in carefully controlled experiments.[3]

Classifying experimental methods in science and engineering

In practice, both engineers and scientists engage in a broad spectrum of experimental strategies. These methods differ in purpose, complexity, size, temporal scale, boundary control and data intensity. The following taxonomy organizes these experimental approaches across disciplines, from the natural and social sciences to economics, engineering, marketing and education. As you review this landscape, keep your own research questions in mind: Which approaches are most likely to illuminate your problem? Which might uncover unexpected insights? Highlight those that align most closely with your needs, and take note of others that may also be worth considering.

DOI: 10.4324/9781003727521-20

Purpose	*Descriptive.* Observe processes without the intention of assessing causal links. *Fundamental.* Identify causal links by exploring the underlying mechanisms. *Evaluative.* Study the effectiveness or impact of specific interventions (e.g., a new medication). *Calibration.* Determine safety factors and develop standards for engineering practice. *Measurement.* Quantify specific properties of materials or systems. *Verification.* Confirm established theories or models under specific conditions. *Identifying hidden mechanisms.* Uncover underlying processes that are not readily observable. *Process exploration.* Understand how a specific process operates under various conditions. *Variable examination.* Assess how one variable changes in relation to another. *Inductive identification of laws.* Gather broad datasets through systematic experimentation across a range of conditions. The resulting data are often plotted in dimensionless form to help reveal underlying scaling laws.[4]
Reference to theory, model or hypothesis	*Discovery/exploratory experiments.* Used in the early stages of research to investigate unknown phenomena and uncover new insights without a predefined initial hypothesis. A classic example is Gibson's cookie-cutter study of interactive perception. *Hypothesis testing.* Designed to validate specific hypotheses. High-stakes experiments, such as those conducted in supercolliders, fall within this category. *Verification/confirmatory experiments.* Aimed to test and confirm the validity of a theory or model under controlled conditions. A landmark example is the 1919 solar-eclipse experiment that confirmed Einstein's theory of relativity by observing the bending of light around the sun. *Extension experiments.* Conducted to test conditions beyond the known limits of a theory's validity. Unexpected results may challenge existing assumptions and compel theorists to revise their models.
Project maturity	*Pilot/feasibility experiments.* Early-stage studies designed to explore a process and its key variables. With a limited scope and simple setups, they highlight qualitative trends and guide future research. *Full experimental studies.* Comprehensive investigations that rigorously test hypotheses, validate models and generate quantitative data to provide definitive conclusions.

(*Continued*)

Physical complexity and size	*Reductionism.* Focuses on individual components, such as molecules in materials, neural activity in cognition or a welded joint in an airplane wing. This approach enables precise measurement and robust analysis, but some phenomena depend on interactions and emergent behavior may not be predictable. *Holistic experiments.* Examine the entire system to capture emergent and collective behaviors. These experiments provide broader insights but are often more difficult to quantify and analyze. *Reduced-size experiments.* Small-scale, cost-effective tests such as centrifuge modeling, microfluidics or wind tunnel experiments allow controlled studies while minimizing resources. *Enlarged-size experiments.* Tests designed to enhance the visibility of specific processes and capture finer details, such as observing particle aggregation under microgravity conditions. *Full-scale experiments.* Maintain real-world dimensions and boundary conditions, avoiding scaling issues and providing the most realistic representation of the system.
Engagement with the subject under study	*Active methods.* Energy is introduced to elicit the system's response to specific stimuli (e.g., ultrasonic testing, thermistors, stressing a material to observe deformation). *Passive methods.* The system is monitored without interference so that measurements capture phenomena in their natural state (e.g., stethoscope, acoustic emission and thermocouples). *Sample-based with lab-controlled conditions.* These experiments require careful sampling protocols and randomized test sequences to avoid biases.[5] *Serendipitous discovery.*[6] Many breakthroughs arise from unexpected observations, accidental discoveries and sheer curiosity.
Type of observations	*Lagrangian measurements.* Observation points move with the process, such as tracking particles or tagged molecules, inferring travel paths from cellphone data, or capsule endoscopy, in which a small camera migrates along the gastrointestinal tract. *Eulerian measurements* (invasive or boundary-based). Observation points are fixed in space. This is the most common experimental design because of its ease of implementation, economy and compatibility with the analysis of spatial fields. Sensors may be installed internally, such as invasive glucose monitors, or used to assess system boundaries, such as thermal imaging devices or strain gauges.

(*Continued*)

	Indirect measurements. The target parameter is inferred through related measurements, introducing some uncertainty due to assumptions or correlations. *Destructive testing.* The specimen is altered or consumed, providing critical data but at the cost of specimen loss (e.g., material strength tests). *Non-destructive testing.* Allows repeated measurements and process monitoring without damaging the specimen, using methods like X-rays or ultrasound.
Variables and data intensity[7]	*Single-variable monitoring: minimal instrumentation but many tests.* This approach examines one variable at a time to isolate its effects under controlled conditions, as in Mendel's genetic studies. Typically, it involves low-cost setups. While effective, single-variable tracking requires numerous tests and can lead to a combinatorial explosion. *Data-intensive experiments.* Employ multiple sensors to collect vast amounts of data from a small number of highly instrumented tests. These experiments may also involve multi-physics sensing to observe interacting processes and uncover coupled phenomena and interdependencies.
Timing and test sequencing	*Cross-sectional – Snapshot.* Captures a single moment in time across multiple subjects or ongoing tests, providing a quick and cost-effective view of a phenomenon. However, it is limited in establishing causality. Example: assessing working memory in individuals of different ages simultaneously to explore age-related differences. *Longitudinal experiments.* Track changes over time for a single subject or group, revealing specimen-specific trends. While they provide rich temporal insights, they may lack statistical relevance for the broader population. Example: monitoring the evolution of working memory throughout the life of a single person. *Same sample, multiple successive stimuli.* A single sample is subjected to different stimuli in sequence. This approach is feasible when the system is linear and time-invariant, ensuring that the order of stimulus application does not affect the outcome. *Different samples, different stimuli.* Each sample is exposed to a unique stimulus, preventing cumulative effects, especially in nonlinear, time-varying systems.

Which approach should you take?

The answer is not always obvious, but it is the one that brings you closer to genuine understanding. The chosen method must suit the problem, but it will also reflect who you are: your skills, your mindset and the tools available to you. Each of us brings a unique experimental style, shaped by our background and training. However, science and engineering require more. To grow as researchers, we must move beyond our comfort zones, expand our toolkit, explore new techniques and refine our ability to ask the right questions. The most effective investigators are not only methodical; they are also curious, adaptable and resolute amid complexity.

Exercises

1 Explore five significant experiments in history. Study them and describe their approach using the taxonomy above. Consider experiments such as Newton's study of light composition using a prism, the Laser-Interferometer Gravitational-Wave Observatory (LIGO), Millikan's oil droplet experiment, Beaumont's study of human digestion, or Tversky and Kahneman's studies that led to "prospect theory".
2 Identify and describe key experiments that have shaped your field of interest.

Notes

1 As paraphrased from French physiologist C. Bernard (1813–1878): *The experimenter must act as the interpreter of nature, designing experiments so rigorously that nature cannot mislead him.*
2 See essay on *Knowledge Creation Throughout History (Section 1).*
3 See essay on *Hidden Magic in Science and Engineering (Section 2).*
4 See essay on *The Dimensional and the Dimensionless (Section 3).*
5 See essay on *Statistical Concepts for Experimental Design and Execution (Section 3).*
6 See essay on *Life Choices and Chances (Section 2).*
7 See essay on *Disruptive Technologies: Rethinking Methodologies and Shifting Paradigms (Section 3).*

Essay 18

DISRUPTIVE TECHNOLOGIES

Rethinking methodologies and shifting paradigms

New tools change the questions we can ask and the way we inquire

Technological advancements have historically served as powerful catalysts for transformative shifts in engineering and science. Consider the humble lens. First conceived in the 13th century, it led to two groundbreaking inventions that revolutionized our understanding of the world at opposite ends of the scale. The telescope enabled G. Galilei (Italy, 1564–1642) to observe Jupiter's moons, challenging the geocentric view of the universe and sparking a revolution in astronomy. Meanwhile, the microscope allowed A. van Leeuwenhoek (Holland, 1632–1723) to observe bacteria for the first time, which ignited the biology revolution.

Fast forward to the 1980s: the development of scanning tunneling microscopy and atomic force microscopy[1] helped launch the nanotechnology revolution. Today, the extraordinary convergence of microelectronics, sensors and computing power invites us to rethink experimental research once again. The following example illustrates how these disruptive technologies can drive profound methodological shifts.

From "many simple tests" to "a few information-rich tests"

Traditional laboratory "element" tests are designed under the assumption that each test represents conditions at a single point in the system under study. This implies that the relevant fields (e.g., stress, strain, temperature, fluid pressure and chemical composition) are homogeneous within the

DOI: 10.4324/9781003727521-21

specimen. These tests control one variable at a time, involve minimal instrumentation, require uniform boundary conditions and rely on uncomplicated models for data interpretation; however, multiple tests are necessary to fully understand the system as a whole.

Recent technological advances have enabled an alternative approach: conducting a single, highly instrumented experiment under complex, non-homogeneous boundary conditions, using distributed multi-physics sensors. In such settings, data interpretation typically depends on numerical modeling within the framework of inverse problem solving. More recently, however, these rich data streams can be fed directly into machine learning algorithms or used to develop adaptive digital twins... What paradigm shifts might emerge from these changes in experimental research methods?

	Traditional	*Post-microelectronics era*
Philosophy	Many simple element tests	A few information-rich tests
Boundary conditions	Elementary, homogeneous	Complex, well-monitored
Field conditions	Homogeneous fields	Heterogeneous fields
Instrumentation	Minimal, low-cost	Extensive, multi-sensor, multi-physics
Measurements	Very few	Numerous, spatial/ temporal, multi-physics
Interpretation	Uncomplicated models	Inverse problem or AI/ML
Information per test	Very limited	As much as needed
Number of tests, time demand	Many tests, long campaigns	One may be sufficient, shorter campaigns
Subjacent advantage	Transparency	Lower risk of confounding variables
Researcher-specimen interaction	More intimate	Typically, more detached

Essential takeaway

Throughout the history of science and engineering, technological breakthroughs have consistently advanced experimental research, occasionally prompting major shifts in methodology. Engage with these developments and contribute to the next wave of scientific progress enabled by technological innovation.

Exercises

Consider the study of thermoelastic coupling in a solid material of your choice.

- Simple element test: Heat a freely standing beam to a preset temperature and measure its change in length. Repeat the test for a wide range of temperatures to explore nonlinear effects up to a maximum temperature T_{max}.
- Information-rich test: Expose one end of the beam to T_{max} and monitor the displacement field using photography and digital image correlation. Simultaneously, capture the evolving temperature field with an infrared camera.

Discuss the test requirements and data interpretation strategies for each case. What insights can be gained from one test that are not accessible through the other? What are the advantages and limitations of each approach?

Note

1 See essay on *Advanced Characterization Systems (Section 4).*

Essay 19

STATISTICAL CONCEPTS FOR EXPERIMENTAL DESIGN AND EXECUTION

Statistics is the grammar of science[1]

The only certainty in experimental research is its pervasive uncertainty! Statistics plays a vital role in managing this uncertainty: it guides experimental design and execution (discussed in this essay) and provides essential tools for data analysis and model development (addressed later in Section 5).

Experimental program: controlling combinatorial explosion

Consider a system whose response depends on multiple independent variables and where the effect of changing one variable may vary depending on the levels of other variables. In such cases, the experimental program must explore all variables and their interactions across different levels. The goal is to design an efficient test program that yields maximal insight while minimizing bias:

- *Full factorial design.* Tests are conducted for every level of every variable. This approach rapidly leads to a large number of tests through "combinatorial explosion". For example, consider $n = 4$ independent variables with $\lambda = 5$ levels each: the full test program would involve λ^n tests, $5^4 = 625$ in this case.[2]
- *Program reduction by hierarchical sequencing.* This adaptive approach starts with a small experimental program, conducting tests in a planned sequence to gradually refine hypotheses and conditions based on earlier results. It can help identify the most important controlling variables and reveal subtle effects or relationships that might not have been initially

DOI: 10.4324/9781003727521-22

apparent. Rigorous assessment at every stage is essential to prevent the accumulation and propagation of experimental bias.

- *Program reduction using Taguchi orthogonal arrays.* This approach selects a carefully structured subset of experiments in which each level of every variable appears the same number of times and all pairs of variable levels are equally represented. The objective is to maximize the overall signal-to-noise ratio, favoring efficiency over resolution. Taguchi test matrices are readily available online; in the example above, with 4 independent variables with 5 levels each, the required number of tests is reduced to 25.
- *Factorial design with blocking.* When an experimental program may be affected by a nuisance variable, experiments can be arranged into properly selected blocks to overcome its effects.

Experimental program: execution strategy

Experiments are affected by a range of unwanted variables and effects that occur alongside the study. This uncertainty is not chaotic but structured: it can be decomposed, understood and managed. Effective mitigation depends on the nature of these uninvited intruders[3]:

- *Nuisance variables* are not part of the primary set of independent variables but affect the response and add unwanted variability. Examples include operators with different skills and attitudes, time-dependent fatigue or boredom and air-conditioning cycles. Mitigation efforts focus on test blocking and randomization.
- *Confounding variables* vary systematically with both independent variables and the outcome, which makes it impossible to distinguish their relative effects. This can lead to incorrect conclusions about causality. For example, a study on the effect of cold weather on school absenteeism might be skewed by the seasonal flu, which acts as a confounding variable. Common control strategies include test sequence randomization and extending the experimental program to explicitly include the confounding variable.
- *Random errors* have a zero mean and tend to cancel out on average. They define the measurement precision (or variance) and are often associated with equipment limitations, sample variability and procedural inconsistencies. At a given test point, such errors are mitigated through repeated measurements.

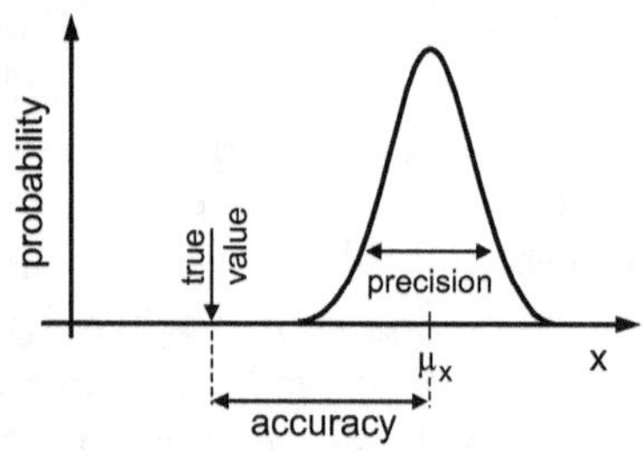

- *Systematic errors* consistently skew results in a particular direction, affecting the measurement accuracy (or bias). They can originate from sample selection bias, measurement bias (e.g., equipment wear, sensor drift, creep, evolving reagent quality, changing population characteristics), and interpretation bias. They are controlled through blinding (participants and researchers), meticulous experimental design, standardized protocols, frequent calibration and rigorous controls. Measurement biases may be corrected during post-processing using calibration data.

Blinding involves withholding information from participants and researchers during the study to avoid the biasing effects of expectations, beliefs and preconceived notions on the results. While particularly relevant in fields such as medicine and social sciences, blinding offers advantages in nearly all experimental programs. Researchers should always establish the "zero-effect baseline" before applying an intervention or changing independent variables and use control groups to assess placebo effects in the absence of intervention. The system's response relative to the baseline quantifies the actual impact.

Other statistical guidelines - Replicates

Statistics also help determine the number of tests or replicates required to ensure adequate statistical power. When random errors dominate, more tests are required to detect smaller effects with higher confidence, especially in noisier systems. In fact, if we want the mean of the measured values to be within an error E_y of the true value with ~95% confidence, the required number of tests n is[4]

$$n = 4\left(\frac{\sigma}{E_y}\right)^2$$

where σ is the standard deviation of the response (assuming a normal distribution with known variance). For example, to ensure the mean stiffness is within 1 GPa of the true value when individual measurements have a $\sigma = 5\,\text{GPa}$ standard deviation, we would need to perform $n = 100$ tests. Alternatively, we could improve the testing procedure to reduce σ (increased precision) or relax the desired precision E_y for the mean.

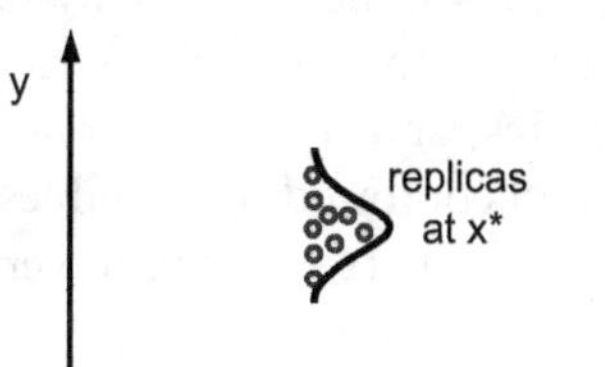

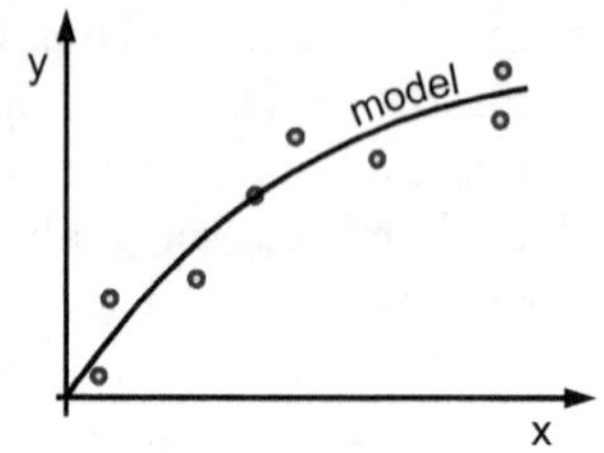

In a full factorial design, the total number of tests is multiplied by the number of replicates at each condition. For example, obtaining 3 replicates at each point for 4 independent variables with 5 levels each would require $3 \times 5^4 = 1{,}875$ tests. Replicates are needed to estimate the local variance of the response y at a given test point x^*. However, in most cases, the goal is to develop models $y(x)$, and we therefore compute the aggregate error relative to the model.[5] Random errors tend to average out across trends, even without replicates. While model fitting takes advantage of the overall trend and error structure, achieving reliable fits still requires a substantial number of tests.

Bottom line

Statistics help us confront the inherent uncertainty in experimental studies. Statistical tools guide the design of efficient test programs that avoid combinatorial overload while preserving statistical power. During test execution, statistical thinking encourages deliberate action: randomizing and blinding test sequences, designing purposeful replication and implementing regular calibration and rigorous procedures to prevent the propagation of bias. By embracing statistical rigor, we improve the quality of our experiments, and both the reliability and scientific value of the data they produce.

Exercises

Choose a system of interest and identify its three most important independent variables. If you plan to test five levels per variable, how many tests are required for a full factorial design? Find a corresponding Taguchi orthogonal array online: how many tests does it require, and do its underlying assumptions (e.g., limited interactions) apply to your system? Will you run replicates, and if so, how many per point and why? Identify potential nuisance and confounding variables, and describe measures to minimize them, reduce random errors, and prevent systematic biases.

Notes

1 K. Pearson (1857–1936), British mathematician and statistician.
2 See reduced dimensionality when posing the problem in dimensionless form – Essay on *Buckingham Pi Theorem: Dimensionless Groups (Section 3).*
3 Review the essay on *Variables (Section 3).*
4 From the central limit theorem.
5 See essay on *Data and Models: Errors and L-Norms (Section 5).*

Essay 20

THE DIMENSIONAL AND THE DIMENSIONLESS

Dimensions frame physical quantities;
dimensionless variables reveal their connections.

The world cannot be captured by numbers alone. Every measurable quantity combines a numerical value with its unit, and these units are not mere labels: they are fundamental components of equations. Examining the dimensions of variables provides critical insights into the behavior of physical systems, from predicting scaling laws to verifying the consistency of models. Let us begin by exploring dimensions and the crucial role they play in deepening our understanding of the physical world.

Dimensions and units

There are seven fundamental physical *dimensions* (measurable properties) and their corresponding *units* (standardized quantities used to express the magnitude of each dimension): mass in kilograms M [kg], length in meters L [m], time in seconds t [s], electric charge in coulombs q [C], amount of substance in moles n [mol], luminous intensity in candelas I [cd] and temperature in kelvins T [K]. While electric current in amperes i [A] is a derived unit, defined as $1A = 1C / s$, it is adopted as a base unit in the International System of Units because it is more practical for studying electrical phenomena. The distinction between variables, dimensions, and units is illustrated next:

DOI: 10.4324/9781003727521-23

Variable	*Dimensions*	*Units*
Velocity	$\frac{L}{T}$	$\frac{\text{m}}{\text{s}}$
Thermal conductivity	$\frac{ML}{t^3 T}$	$\frac{\text{W}}{\text{mK}} = \frac{\text{kg m}}{\text{s}^3\text{K}}$

An equation that is not dimensionally homogeneous is not physically correct

Always check the dimensional homogeneity of equations, typically by verifying the consistency of their units. For example, position x [m], velocity v [m/s] and acceleration a [m/s²] are related through the following dimensionally homogeneous equation:

$$
\begin{array}{cccccccc}
x = & x_o & + & v & \Delta t & + & \frac{1}{2} & a \ \Delta t^2 \\
\hline
[\text{m}] & [\text{m}] & & \left[\frac{\text{m}}{\text{s}}\right] & [\text{s}] & & & \left[\frac{\text{m}}{\text{s}^2}\right]\left[\text{s}^2\right]
\end{array}
$$

Dimensional homogeneity can even help us recall the correct expression for a physical process. Consider diffusion: since the diffusion coefficient is D [m²/s], the correct diffusion equation must be

$$\text{wrong: } \frac{\partial^2 c}{\partial x^2} = D\frac{\partial c}{\partial t} \qquad \text{correct: } D\frac{\partial^2 c}{\partial x^2} = \frac{\partial c}{\partial t}$$

Adding qualifiers to units provides additional insight. For instance, how can we relate the bacteria count measured in one milliliter of extracted water cc_{fluid} [cells/ml] to the bacteria count in the host sediment volume cc_{sed} [cells/ml]? Let's add qualifiers to both volumes:

$$cc_{sed}\left[\frac{\text{cells}}{\text{ml}_{sed}}\right] = \langle ? \rangle \times cc_{fluid}\left[\frac{\text{cells}}{\text{ml}_{fluid}}\right]$$

Once the unit "milliliter" is qualified, dimensional homogeneity reveals that the missing factor $\langle ? \rangle$ must have units of $[\text{ml}_{fluid} / \text{ml}_{sed}]$. Clearly, this dimensionless factor is the porosity $n = \text{ml}_{fluid} / \text{ml}_{sed}$, so that $cc_{sed} = n \times cc_{fluid}$. The meaning of a quantity depends on its dimensional context.

There is no information about dimensionless numbers

Consider the specific surface area S_s [m^{-1}] of a cube, defined as surface area [m^2] per unit volume [m^3]. From a dimensional standpoint, we can anticipate that the specific surface is inversely proportional to the cube's characteristic length, $S_s \propto L^{-1}$, but we gain no insight into any missing dimensionless factor. Indeed, since a cube has six faces, the specific surface area is exactly $S_s = 6L^2 / L^3 = 6 / L$. Similarly, dimensional homogeneity tells us that the period of a pendulum T [s] of length ℓ [m] in a gravitational field with acceleration g [m/s^2] must scale as $T \propto \sqrt{\ell / g}$. However, determining the exact physical relationship $T = 2\pi\sqrt{\ell / g}$ requires proper physical analysis to identify the missing dimensionless factor 2π.[1]

Transcendental functions require dimensionless arguments

The arguments of exponential, trigonometric, logarithmic and Bessel functions (among others) must be dimensionless. Here are three common examples:

- The particle motion during wave propagation is $u = A\sin(\kappa x - \omega t)$, where the wave number $\kappa = 2\pi / \lambda$ is inversely proportional to the wavelength λ [m], and the angular frequency $\omega = 2\pi / T$ is inversely proportional to the period T [s]. Hence, the argument of the sine function is dimensionless.
- A wave propagating through a homogeneous medium experiences exponential decay $e^{-x/\beta}$, where the "skin depth" β [m] is the distance at which the wave amplitude has decayed by a factor of $1/e$. Consequently, the argument x / β is dimensionless.
- Sound intensity I [W/m^2] spans many orders of magnitude, so it is better assessed on a logarithmic scale to capture both very low and very high intensities. The decibel is defined as $\text{dB} = 10\log_{10}(I / I_0)$ where the typical normalizing reference $I_0 = 10^{-12}$ W/m^2 corresponds to the threshold of human hearing. Clearly, the ratio I / I_0 is dimensionless.

To emphasize: the logarithm of a dimensional number does not exist! The value must be first converted into a dimensionless quantity, and the chosen normalization must be stated explicitly. For example, given a stress level $\sigma = 10$ MPa, its logarithm could be written as $\log(\sigma / \text{MPa}) = 1$ or $\log(\sigma / \text{kPa}) = 4$ depending on the value adopted for normalization.

Normalization: comparing data and identifying underlying trends

Consider the proportions of the human body. There is a marked variation in height: 95% of females are between 1.46 m and 1.76 m, and 95% of males are between 1.55 m and 1.85 m. Similarly, there is significant variation in arm span (fingertip to fingertip). However, the ratio of arm span to height is surprisingly close to 1:1 across all cases, typically ranging between 0.98 and 1.04. In this case, normalized measurements collapse into a narrow value, revealing a universal proportion of the human body.

A similar approach applies to stress–strain data across different materials. Strain, defined as the ratio of deformation to the initial length, is inherently normalized and therefore requires no further scaling. Stress, on the other hand, can be normalized by each material's peak strength to enable more meaningful comparisons of material response.

In general, normalization enables comparison of experimental data from diverse sources within the same dimensionless framework. Sometimes, all data will converge onto a common trend; when this happens, we are on the path to uncovering the physical controls governing the phenomenon.

Dimensional homogeneity across scales: macroscale ↔ microscale

Dimensional analysis helps reveal connections between macroscopic responses and the underlying microscale processes. For example, consider a material exhibiting time-dependent macroscopic behavior, such as creep: there must be internal microscale processes that also depend on time, such as reaction rates, diffusion, viscosity, flow velocity or inertial forces. The converse is also true: if a dimension is absent at the microscale, it cannot emerge at the macroscale. For instance, a granular packing composed of infinitely rigid particles cannot exhibit stress-dependent macroscopic behavior, since stress has no physical meaning at the particle level.

Summary

The analysis of dimensions is surprisingly powerful! It helps uncover inconsistencies and missing variables, hints at hidden relationships and underlying causes, and reduces the apparent complexity of a problem. So here is one final, "foundational" recommendation before we move on to the next essay on dimensional analysis: always include units when introducing variables and throughout all physical analyses and computations. It will save you time t [s]!

Exercises

Consider the following well-known equations. What are the appropriate units for the missing quantity ⟨?⟩ in each case?

Newton's law of gravitation	$F = \langle ? \rangle \frac{m_1 m_2}{r^2}$	$\langle ? \rangle$	$\left[\frac{\quad}{\quad}\right]$
Coulomb's law of electrostatics	$F = \langle ? \rangle \frac{q_1 q_2}{r^2}$	$\langle ? \rangle$	$\left[\frac{\quad}{\quad}\right]$
Ideal gas law	$pV = \langle ? \rangle nT$	$\langle ? \rangle$	$\left[\frac{\quad}{\quad}\right]$
Hooke's law	$F = \langle ? \rangle \delta$	$\langle ? \rangle$	$\left[\frac{\quad}{\quad}\right]$
Wave equation	$\frac{\partial^2 u}{\partial t^2} = \langle ? \rangle \frac{\partial^2 u}{\partial x^2}$	$\langle ? \rangle$	$\left[\frac{\quad}{\quad}\right]$

Note: F = force; m = mass; r = distance; q = electric charge; p = pressure; V = volume; n = amount of substance; T = temperature; δ = deformation; u = particle motion; t = time; x = position in space.

Note

1 See essay on *Enduring Simple Machines: From Archimedes to Galileo and Pascal (Section 4)*.

Essay 21

BUCKINGHAM PI THEOREM

Dimensionless groups

The mountains flowed before the Lord[1]

Any physical system can be described using a set of variables. Each carries a physical dimension and a corresponding unit that provides scale[2]: mass [kg], length [m], time [s] or electric charge [C]. These dimensional variables are the scaffolding of physical thought. Yet, the same system can often be more powerfully expressed using dimensionless groups, i.e., combinations of variables that transcend units and reveal the controlling relationships within the phenomenon.

> **BUCKINGHAM PI THEOREM[3,4]:**
>
> If a physical system involves n dimensional variables $f\left(v_1, v_2, \ldots v_n\right)$ expressed in terms of k dimensions, then the system can be described by $n-k$ dimensionless Π-groups $F\left(\Pi_1, \Pi_2, \ldots \Pi_{n-k}\right)$

This theorem has two immediate advantages. First, it suggests that two physical systems may appear different in terms of the variables v_i, yet they will exhibit a similar response if their corresponding dimensionless groups Π_j are the same. For example, the tipping stability of a rectangular box depends on its height h [m] but also on its width B [m]. In fact, the ratio h/B best describes the overturning stability of a prismatic structure, from a high-rise building to a matchbox – this is similarity!

DOI: 10.4324/9781003727521-24

Second, expressing a problem in dimensionless form reduces its effective dimensionality: instead of exploring the problem in an n-dimensional space, we can work within a smaller $n-k$ space defined by dimensionless groups. This significantly reduces the combinatorial explosion in the number of required experiments (or simulations): testing n variables at λ levels would require λ^n tests[5]; in contrast, exploring the $n-k$ dimensionless groups also at λ levels would require only λ^{n-k} tests. Hence, the total effort is reduced by a factor of λ^k. To illustrate: if a system is described by $n = 5$ variables in terms of $k = 3$ fundamental dimensions, and each variable is explored at $\lambda = 4$ levels, a full factorial design would require $4^5 = 1024$ tests; however, reformulating the problem in terms of the $n-k = 2$ dimensionless Π groups reduces the requirement to just $4^2 = 16$ tests, a dramatic improvement made possible by normalization.

Implementation

To apply the Buckingham Pi Theorem, let your physical insight guide you through the following four steps:

1. Identify the *main n-variables* involved in the phenomenon or process under study.
2. List the *dimensions* for each variable, i.e., k-dimensions.
3. Combine the variables to create $n-k$ independent *dimensionless* Π*-groups*.[6]
4. Identify the *physical meaning* of each dimensionless group and its implications for the system's behavior.

For example, in the case of the overturning stability of the prismatic structure, the ratio h / B is the body "slenderness" and plays a central role in the balance between the overturning moment and the resisting moment, i.e., equilibrium.

A few additional considerations help us fully exploit the power of the Buckingham PI theorem.

- If only one of the listed variables carries a particular dimension, it implies that a relevant variable is missing from the formulation...
- Ensure that all the n-variables identified in the first step are involved in the final n-k independent dimensionless Π-groups.
- Multiplication $\Pi_i \times \Pi_j$, division Π_i / Π_j and exponentiation Π_j^b preserve the dimensionless form. Use these operations to combine dimensionless groups and construct a final set with clear physical meaning.

Illustrative examples

Example: A pendulum in vacuum

A straightforward experiment quickly reveals that the cord length $\ell\,[\mathrm{m}]$ plays a critical role in determining the pendulum's period $T\,[\mathrm{s}]$. What about the mass of the bob, a common intuitive guess? If we include mass as a variable, dimensional consistency requires at least one other variable involving mass; but in a vacuum, no such variable exists, so we can conclude that the bob's mass is not a governing parameter. However, we are still missing a key variable needed to balance the dimensions of length and time in $\ell\,[\mathrm{m}]$ and $T\,[\mathrm{s}]$. That variable is gravity g [m/s^2]. Consequently:

Parameter	*Symbol*	*Unit*	*Dimension*
Period	T	s	T
Cord length	ℓ	m	L
Gravity	g	$\mathrm{m/s^2}$	$\mathrm{LT^{-2}}$

$n = 3$ variables, $k = 2$ dimensions

So, there is only $n - k = 1$ dimensionless group $\Pi = T^2 g/\ell$. The formal physical solution derived from energy conservation predicts that $T = 2\pi\sqrt{(\ell/g)}$. While dimensional analysis cannot produce the numerical factor 2π, it successfully captures the fundamental relationship between the period and the square root of the cord length.[7]

Example: Liquid flow through a pipe

What variables govern the flow rate of a liquid through a cylindrical pipe? The key variables are readily apparent:

Parameter	*Symbol*	*Unit*	*Dimension*
Pressure difference	Δp	Pa	$\mathrm{ML^{-1}T^{-2}}$
Liquid viscosity	μ	$\mathrm{Pa\cdot s}$	$\mathrm{ML^{-1}T^{-1}}$
Pipe length	ℓ	m	L
Flow rate	q	$\mathrm{m^3/s}$	$\mathrm{L^3T^{-1}}$
Tube radius	r	m	L

$n = 5$ variables, $k = 3$ dimensions

There are $n = 5$ variables in terms of $k = 3$ dimensions: length, time and mass. Therefore, the problem of flow through pipes can be stated in terms of $n - k = 2$ dimensionless groups. For example, we can create $\Pi_1 = \ell/\mathrm{r}$ which defines the pipe's geometric slenderness and $\Pi_2 = \Delta p r^3/(\mu q)$. Alternatively, we could also select $\Pi_2/\Pi_1 = \Delta p r^4/(\mu q \ell)$ which hints at the ratio between the pressure-induced driving force and the viscous resistance along the pipe. In fact, the formal mathematical solution for liquid flow through a pipe is $q = (\pi / 8)\Delta p r^4 / (\mu \ell)$ (Hagen-Poiseuille equation). Once again, dimensional analysis cannot produce the $\pi / 8$ factor, but it properly captures the interplay between all physical variables.

Example: Advective-diffusive transport

What are the main variables that define the time scale for advective and diffusive transport across a medium?

Parameter	*Symbol*	*Unit*	*Dimension*
Observation time	t	s	T
System length	ℓ	m	L
Flow velocity	v	m / s	LT^{-1}
Diffusion coefficient	D	m^2 / s	L^2T^{-1}

$n = 5$ variables, $k = 2$ dimensions

We can create $n - k = 3$ dimensionless groups. The group $\Pi_1 = tv / \ell$ compares the observation time to the characteristic time for advection $t_{adv} = \ell / v$. The group $\Pi_2 = tD / \ell^2$ compares the observation time to the characteristic time for diffusion $t_{dif} = \ell^2 / D$. Finally, the group $\Pi_3 = t_{dif} / t_{adv} = v\ell / D$ is the Péclet number, which compares the two characteristic process times: diffusive transport dominates when $Pe \ll 1$, while advective transport dominates when $Pe \gg 1$.

Some well-known dimensionless groups

The following dimensionless groups are used to compare dominant processes in coupled-physics systems:

Reynolds number	Re	inertial forces/viscous forces (laminar vs. turbulent flow)
Froude number	Fr	inertial forces/gravitational forces (in gravity-driven flow)
Grashof number	Gr	buoyant forces/viscous forces (onset of convective flow)
Capillary number	Ca	viscous forces/capillary forces (microfluidics)
Weber number	We	inertial forces/surface tension forces (droplet breakup)
Bond number	Bo	gravity/surface tension (droplet shape under gravity)
Péclet number	Pe	advective transport/diffusive transport
Damköhler number	Da	reaction rate/advective transport rate
Knudsen number	Kn	molecular mean free path/pore diameter
Brillouin number	Br	wavelength/internal length scale (equivalent continuum vs. ballistic)
Relative stiffness	Rs	ratio between the stiffness of two coupled mechanical elements[8]
Timoshenko number	To	flexural stiffness/inertial effects

Why Π-groups matter

The insights and examples above highlight the powerful advantages of using the Buckingham Pi Theorem to represent physical phenomena through dimensionless groups:

- Helps identify missing or overlooked governing variables.
- Enables consistent comparisons across experiments and systems.
- Reduces dimensionality, dramatically lowering experimental effort.
- Provides straightforward scaling estimates.
- Reveals the dominant processes in coupled-physics systems.

So, the next time you encounter a complex physical problem, consider exploring it in the dimensionless space, where Π-groups lead the way to deeper insight.

Exercises

1. Extend the pendulum example to consider the medium as a viscous fluid with mass. Then, search for analytical solutions. Compare and discuss.
2. Analyze the phenomenon of fluid flow through a porous medium. List all variables that you consider most relevant to the phenomenon. Create dimensionless groups and carefully study and discuss the physical significance of each Π-group.
3. Analyze a problem within your field of research using dimensional analysis. Follow the four steps outlined above.
4. Extend the use of dimensionless ratios to (a) numerical simulations and (b) machine learning. Discuss how dimensionless analysis can be implemented in each context, and explore the potential advantages it offers in terms of generalization, efficiency and insight.

Notes

1. Song of Deborah, Book of Judges (Old Testament). In physics, Deborah number De is the ratio between the viscoelastic relaxation time and the observation time.
2. See essay on *The Dimensional and the Dimensionless (Section 3).*
3. E. Buckingham (1914). On physically similar systems; illustrations of the use of dimensional equations. *Physical Review*, 4, 345–376.
4. For a detailed discussion, see P.W. Bridgman (1922). *Dimensional Analysis*. Yale University Press.
5. See essay on *Statistical Concepts for Experimental Design and Execution (Section 3).*
6. Matrix algebra can be used, but it is not necessary in most cases. See J.C. Santamarina and D. Fratta (2005). *Discrete Signals and Inverse Problems. An Introduction for Engineers and Scientists.* John Wiley & Sons Ltd.
7. See essay on *Enduring Simple Machines: From Archimedes to Galileo and Pascal Section 4).*
8. Examples: cantilever beam connected to a spring; bone and hip replacement; fiber-reinforced material.

Essay 22

SIMILARITY AND SCALING LAWS

A giant made [of] the same proportion of limbs that is found in an ordinary man [with the same bone hardness and strength] he will fall and be crushed under his own weight... The smaller the body, the greater its relative strength[1]

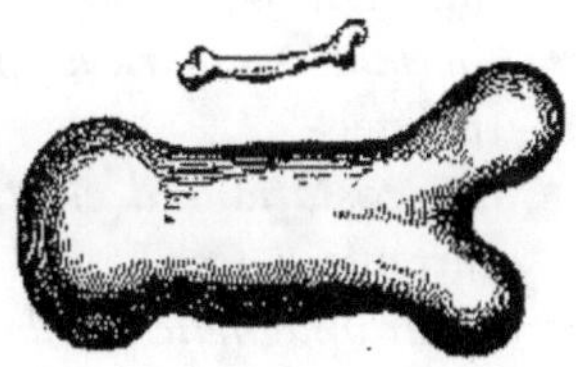

Let us begin by understanding Galileo's remarkable insight. Imagine a cube of size $h[\mathrm{m}]$, made of a material with unit weight $\gamma\left[\mathrm{N/m^3}\right]$. Its volume is h^3, so its weight is $\gamma\ \mathrm{h}^3$. The contact stress at the base, defined as weight over area, is $\sigma_{\mathrm{small}} = \gamma\ h$.

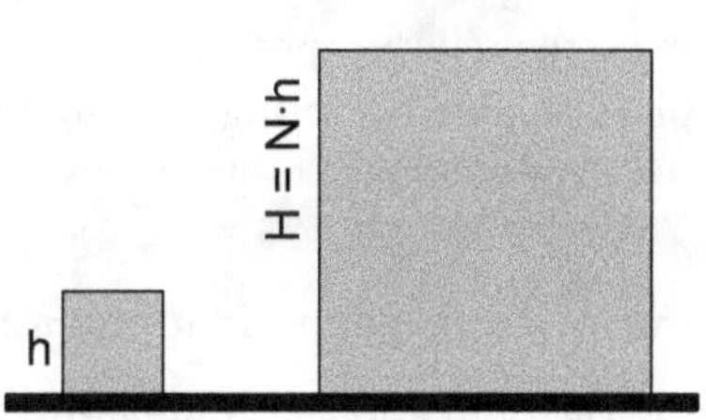

Now consider a geometrically identical cube made of the same material but N times larger, with side length $H = N\ h$. The contact stress at the base of the larger cube is $\sigma_{\mathrm{large}} = \gamma\ H = \gamma\ N\ h$, which is N times larger than of the small cube. Therefore, if the small cube is designed with a reasonable factor of safety, the much larger cube will certainly exceed the material's strength.

Since the human femur has a factor of safety of about 3, a giant ten times larger than us would be crushed under its own weight. This example illustrates that geometric similarity alone does not ensure structural similarity. Successful scaling requires proportional adjustments in material properties, such as increasing strength or reducing density, to preserve functional integrity.

DOI: 10.4324/9781003727521-25

Similarity

Systems are considered "similar" when they exhibit comparable behaviors in terms of dimensionless ratios, despite differences in scale, as per the Buckingham Pi Theorem.[2] To achieve similarity, systems must satisfy:

- *geometric similarity*: angles are equal and all dimensions scale proportionally (same geometric factor)
- *kinematic similarity*: proportional displacements, velocities and accelerations
- *dynamic similarity*: proportional forces (inertial, gravitational, viscous, elastic...)
- *mechanical similarity*: proportional stresses and strains
- *wave similarity*: proportional frequency and amplitude
- *biological similarity*: proportional biological processes
- *chemical similarity*: may apply to aspects ranging from atomic structure to reactivity and physical properties (e.g., phase transition, saturation and diffusivity)
- *thermal similarity*: proportional temperature distribution and heat transfer
- *electrostatic and electrodynamic similarity*: proportional potentials and currents
- *electromagnetic similarity*: wave similarity, material interaction with the electromagnetic field

Total similarity is unattainable. Therefore, experimentalists must identify the key governing parameters and design experiments that closely approximate the relevant physics. It is then essential to evaluate the effects of incomplete similarity through sensitivity analyses and remain prepared to adjust the experimental design to ensure the reliability of results.

Intrinsic size-dependent phenomena

Physical phenomena inherently exhibit size-dependent behavior, which profoundly impacts both experimental design and data interpretation. This raises important questions: How fundamental is a property if its value varies with size? And more importantly, what underlying processes cause these phenomena to depend on scale?

Take biological systems. One might expect that an organism's power consumption P [W] scales linearly with its number of cells, thus with its mass m [kg], leading to $P \propto m$. However, the organism's ability to dissipate heat depends on surface area, $P \propto h^2$, where the characteristic length h scales with mass as $h \propto m^{1/3}$; this leads to $P \propto h^2 \propto m^{2/3}$, a thermodynamic constraint to prevent overheating. Which of these scaling laws is correct, $P \propto m$ *or* $P \propto m^{2/3}$? Experimental data exhibit a sublinear relationship between metabolic power consumption $P \propto m^{3/4}$ and the organism's mass.[3] This proportionality reveals an economy of scale: nature has developed efficient fractal networks (such as vascular tree structures) to overcome the limitations imposed by geometric scaling, leading to the ¾ power law in biological systems. In contrast, certain complex systems show the opposite trend: urban metrics like per-capita wages, rate of innovation and even crime scale superlinearly with city size.[4]

Examples of size-dependent material properties include pore-size-dependent diffusive transport (ranging from Fickian diffusion in large pores to Knudsen diffusion in small pores) and particle-size-dependent behavior in granular materials (gravimetric forces $\propto h^3$ control millimeter-size particles, while surface-related electrical forces $\propto h^2$ gain relevance in submicron-size particles).

Size-dependent parameters may also result from internal length scales L_{int} inherent to the material or system under study. In wood, for instance, zooming in gradually reveals annual rings (cm), wood grains (mm), cells (µm) and polymers (nm). Each of these length scales is crucial in determining the overall properties of the material. For accurate measurement of the effective macro-scale properties, the specimen size h must be larger than any internal length scale in the material $h / L_{int} >> 1$. Likewise, in wave propagation studies, the wavelength λ must be longer than the medium's internal length scale $\lambda / L_{int} >> 1$ to determine equivalent continuum properties.

Materials with self-similar fractal structure defy this requirement: they lack a characteristic length scale as their structural patterns repeat across multiple scales. Consequently, the measured properties of fractal materials are specimen-size dependent.

Closing thoughts

Similarity, assessed in the space defined by dimensionless ratios, is a *sine qua non* condition for translating experimental results to the physical reality under investigation. Discrepancies between expected and observed scaling laws compel us to revisit our assumptions and examine more closely the underlying processes at play. To ensure meaningful results, experiments must also satisfy similarity at the boundaries and account for intrinsic internal scales present in the system. Complementary numerical simulations and digital twins offer powerful means to investigate the influence of boundary conditions and internal heterogeneity; once again, numerical methods emerge as indispensable tools in the experimentalist's toolbox.

Exercises

1 Bonsai trees are miniature versions of full-sized trees, but do they respond similarly to external stimuli? Identify key physiological processes in vascular plants (e.g., fluid transport, structural support and energy capture) and the relevant variables (e.g., sap flow rate, vessel diameter, plant height, gravity and fluid viscosity). Use dimensional analysis to group variables into meaningful dimensionless Π-groups. Apply scaling laws, including Murray's law (optimal branching for fluid transport) and da Vinci's law (conservation of cross-sectional area in branching), to compare bonsai and full-sized trees. Discuss expected deviations from similarity and their causes.

2 Consider the following materials: (a) woven fabrics, (b) fiber-reinforced composites, and (c) natural granite. Identify internal structural features and characteristic length scales (e.g., fiber diameter, yarn spacing, grain size). Create a comparative table of mechanical and thermal properties for both individual components and the bulk material. Develop physical models to explain differences between micro- and macro-scale behavior. Estimate the minimum representative specimen size needed for laboratory testing to capture full material behavior.

3 Choose a material of interest and repeat the analysis from Exercise 2.

Notes

1 G. Galilei (1564–1642), Italian physicist, mathematician, astronomer and philosopher – See A. Frova and M. Marenzana (2006). *Thus Spoke Galileo*. Translated by J. McManus. Oxford University Press.

2 See essay on *The Dimensional and the Dimensionless (Section 3)*.

3 Kleiber's law: after M. Kleiber (1893–1976), Swiss biologist.

4 L.M.A. Bettencourt, J. Lobo, D. Helbing, C. Kühnert and G.B. West (2007). Growth, innovation, scaling, and the pace of life in cities. *Proceedings of the National Academy of Sciences (PNAS)*, 104, 7301–7306.

SECTION 4
Devices and sensors

Essay 23

INTRODUCTION TO DEVICES AND SENSORS

Measure what is measurable, and make measurable what is not so[1]

Before the rise of advanced instrumentation, scientists relied on direct observation to uncover the physical world. Archimedes (~250 BC) deduced the principle of buoyancy by stepping into a bath. G. Galilei (~1603) timed balls rolling down inclined planes with a water clock to discern the laws of motion. I. Newton (1660s) demonstrated that white light is composed of many colors using glass prisms. C.A. de Coulomb (~1784) used a torsion balance to show that the electrostatic force between charges varies inversely with the square of the distance. J.W.S. Rayleigh (1870s) explored acoustic resonance by stretching membranes across wooden rings and using sand grains to visualize nodal patterns. A. Pockels (1880s) made foundational discoveries in surface science using nothing more than kitchen utensils, and J.J. Gibson (~1960) conducted hands-on experiments to reveal how humans extract meaningful information from sensory input.

Today's devices arise from the confluence of multiple technological revolutions, pushing new limits across all fields: miniaturization reaching into the quantum realm; exceptional multi-physics sensing far beyond the limits of human perception; ultra-precise actuation and massive data processing supported by vast databases and artificial intelligence. As we rely increasingly on advanced measurement systems, we become detached from the measurement itself, and the physical reality under study exists only alongside its representation in our imagination.

Yet simple physical elements and concepts remain the building blocks, integrated to create extraordinary instruments that advance experimental research. The essays in this section revisit the foundational physical principles

DOI: 10.4324/9781003727521-27

of mechanics and electromagnetism that underpin measurement, enhancing our understanding of how we measure what we measure. The subsequent section addresses complementary mathematical and data-processing concepts that empower modern devices.

Exercises

1. Checking for fever? Analyze the underlying principles of different temperature measurement methods: touch/palpation, glass thermometers, digital thermometers, non-contact infrared thermometers and wearable temperature patches. How does each method work? What are the underlying assumptions and limitations of each approach?
2. See beneath the hood: find images of the engine in a Ford Model A and a modern car. Identify the key components in each case. How many sensors were used in the past compared to today? What does this reveal about the evolution of measurement and control in engineering systems?
3. Our labs look so different… but are they really? Locate images of a laboratory in your field from the early 20th century and from today. Identify the measurement principles and key components in each case. What is genuinely new? To what extent does the saying "what is old is new again" hold true? What are the modern twists?

Note

1 G. Galilei (1564–1642), Italian astronomer and physicist.

Essay 24

ENDURING SIMPLE MACHINES

From Archimedes to Galileo and Pascal

Give me a lever long enough and a fulcrum on which to place it, and I shall move the world[1]

"Classical simple machines" enabled many of humanity's earliest engineering feats. The lever and the wedge date back to prehistory, while the wheel and the inclined plane emerged near the end of the Stone Age (~3000 BCE). More complex mechanisms like the screw and gears appeared by ~350 BCE. At their core, these elemental machines function as "converters": they transform the magnitude and direction of forces and displacements, always involving a trade-off. Far from being mechanical relics, they remain integral to modern devices and sensors, continuing to offer timeless physical insight.[2] In this essay, we revisit these foundational components not as historical curiosities but as enduring tools for design and experimentation.

Basic mechanical elements

Lever

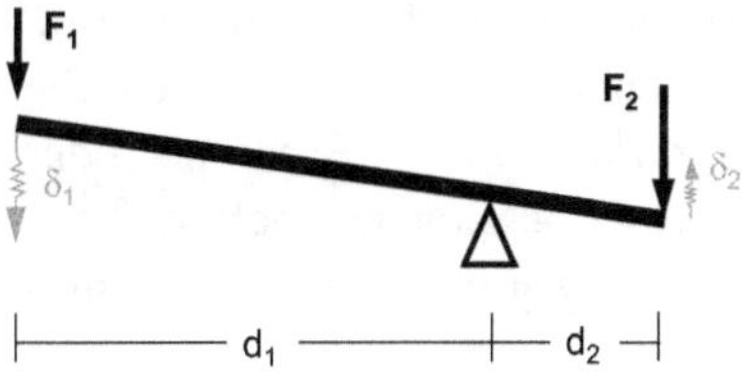

A lever is a rigid bar that pivots around a support point called the fulcrum. From moment equilibrium, the forces on either side of the fulcrum satisfy: $F_1\, d_1 = F_2\, d_2$. A modest force F_1 acting at a longer arm d_1 can produce a larger force F_2 at a shorter distance d_2. This mechanical advantage comes with a corresponding trade-off in displacement: for a rigid beam, $\delta_1 / d_1 = \delta_2 / d_2$. This inverse

DOI: 10.4324/9781003727521-28

relationship is fully consistent with energy conservation and illustrates a recurring theme among all elemental machines.[3]

Inclined plane and wedge

The inclined plane redirects both force and motion between horizontal and vertical directions. Neglecting friction, force equilibrium yields the following relationship between the horizontal H and vertical N forces: $H = N \tan\theta = V(h/\ell)$. A small slope h/ℓ therefore produces a large mechanical advantage. Geometry and energy conservation again reveal the trade-off in displacements $\delta_z / \delta_x = h/\ell$. Force amplification is obtained at the cost of increased travel distance.

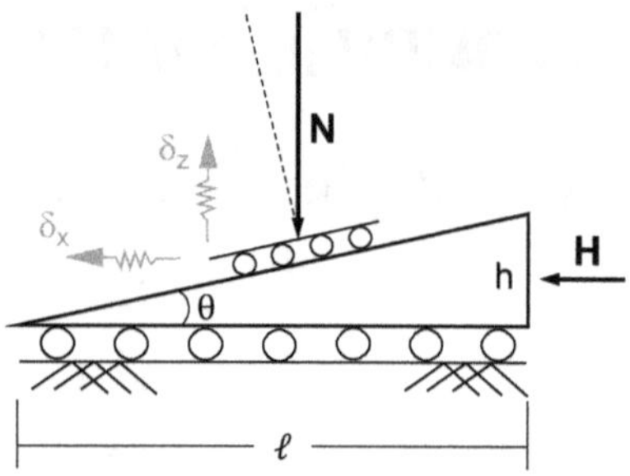

Screws, nuts and washers

A screw is an inclined plane wrapped around a cylinder.[4] The pitch angle α relates the axial advance per turn (or thread pitch h) to the screw perimeter, $\tan\alpha = h/(2\pi r)$. Following the inclined plane analogy and neglecting friction, a twisting force F_t applied at a distance ℓ produces an axial force N along the screw's axis:

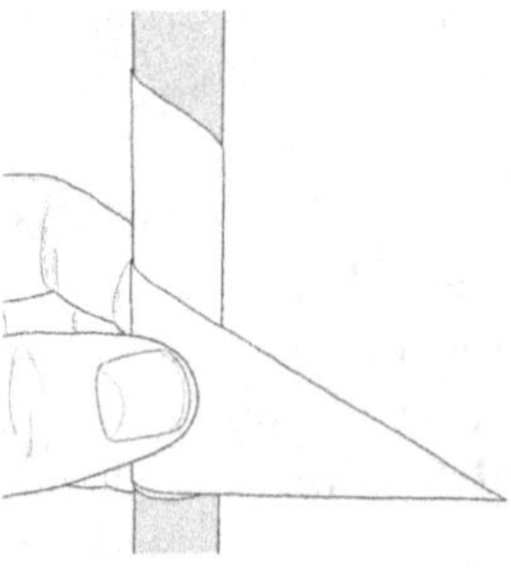

$$F_t \ell = N\, r \tan\alpha \rightarrow N = 2\pi F_t \frac{\ell}{h}$$

Thus, a relatively small torque can generate a very large axial force. Because the ratio l/h is typically large when using a wrench, even modest effort can place a screw under significant tension.

This leads naturally to the question: *how thick should a nut be?* A balanced design ensures that thread shear does not fail before the screw reaches its tensile capacity. Equating axial tension and thread shear:

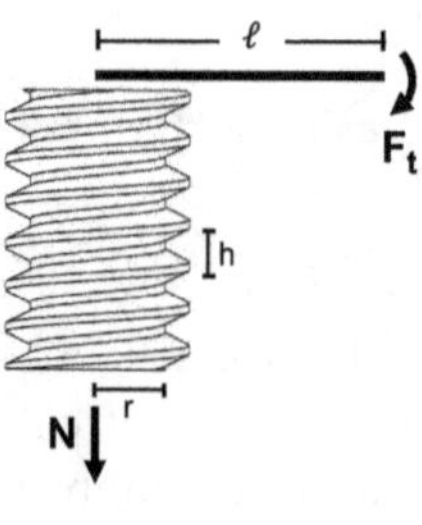

Tension in screw $F_{ten} = \pi r^2 \sigma$ *Screw-nut shear* $F_{sh} = 2\pi r b \tau$

For ductile metals, the shear strength τ is about 60% of the tensile strength σ. This implies that the nut thickness b should be on the order of the screw radius r. In practice, $b \approx 1.6\,r$, ensuring failure occurs preferentially in the bolt rather than by thread stripping.

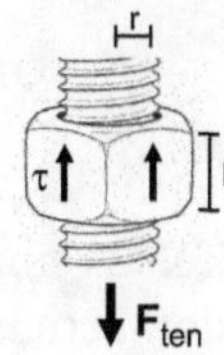

Finally, washers serve several important functions: flat washers reduce friction during tightening, fender washers distribute load over a larger area, sealing washers prevent fluid or gas leakage, and spring washers mitigate loosening due to vibration or thermal cycling.

Screws convert rotational motion into precise linear displacement. For example, a screw with a pitch of $h = 1$ mm advances only 2.8 μm per degree of rotation. This exceptional positional resolution explains their widespread use in micrometers and precision positioning stages.

Wheel

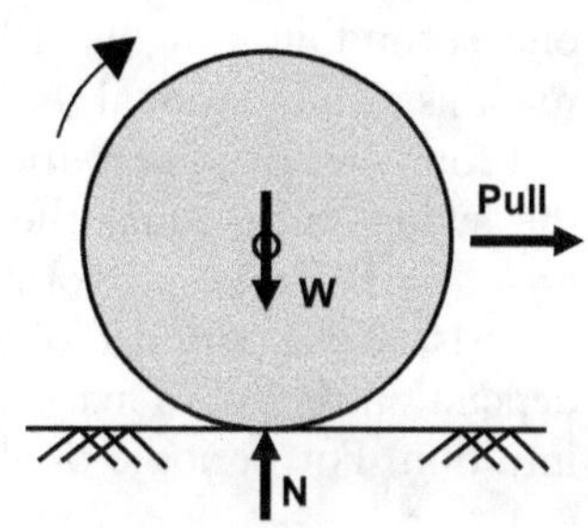

The wheel transformed transportation, agriculture and industry by enabling motion with dramatically reduced resistance. What makes the wheel so special? Clearly, it allows seamless rotation and smooth motion, but its true power lies in minimizing unnecessary resistance. Unlike sliding, rolling greatly reduces friction: when the axle load W is centered above the wheel-ground contact point, virtually no torque is required to maintain motion. The contrast becomes obvious when imagining a square wheel!

Pulley

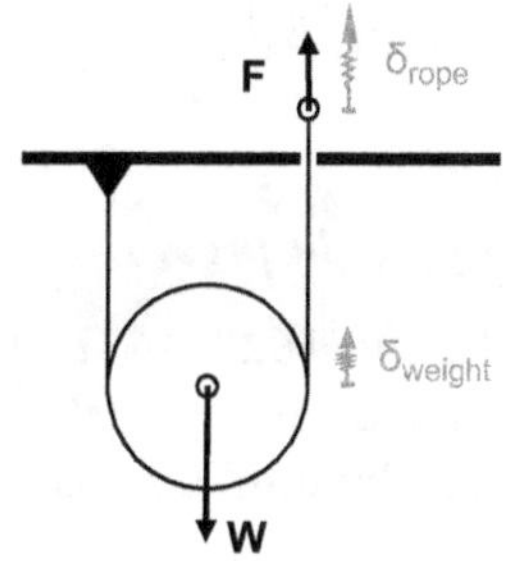

A pulley is a wheel that redirects a force applied to its shaft through a rope wrapped around its rim. In the arrangement shown, force equilibrium gives $F = W/2$. The force F required to lift a weight W is halved, while the displacement doubles: the rope must be pulled twice the distance the load rises, $\delta_{\text{rope}} = 2\delta_{\text{weight}}$. By combining multiple pulleys into compound systems, substantial mechanical advantage can be achieved with the same fundamental trade-off between force and motion.

Gears

Gears enable controlled changes in speed, torque and direction while preserving synchronized motion. Their mechanical advantage is governed by the ratio of teeth numbers N:

$$\textit{angular velocity} \quad \frac{\omega_1}{\omega_2} = \frac{N_2}{N_1} \qquad \textit{torque} \quad \frac{T_1}{T_2} = \frac{N_1}{N_2}$$

Gears remain ubiquitous in modern equipment, from servo motors and precision actuators to micro-electro-mechanical systems (MEMS).

The pendulum

Galileo's study of pendular motion (1564–1642) initiated major advances in timekeeping during the 17th and 18th centuries. Pendulum clocks dominated precision time measurement until the 20th century, achieving accuracies of about one second per day (by comparison, modern atomic clocks lose only one second in roughly 100 million years). Yet the elemental pendulum remains a foundational tool in science and engineering.

From energy conservation, the maximum potential energy at the peak of the swing must equal the maximum kinetic energy at the lowest point, $mgh_{max} = 1/2\, mv_{max}^2$. Notice that the mass m cancels, so the period T of a pendulum does not depend on the bob's mass, contrary to common intuition. For periodic oscillations described by $x(t) = \ell \sin\theta \cos\left(\frac{2\pi}{T} t\right)$ and assuming a small angle θ, the period depends only on the length ℓ and gravitational acceleration g:

$$T = 2\pi\sqrt{\frac{\ell}{g}}$$

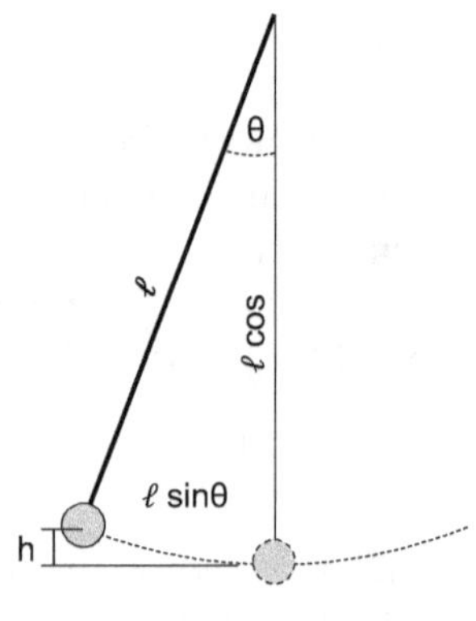

Hydraulic press

Pascal (1623–1662) demonstrated that pressure in an enclosed fluid is transmitted equally in all directions. This principle underlies all hydraulic systems. The hydraulic press, invented by J. Bramah in 1795, is a classic example: a force F_1 acting on a piston of area A_1 produces pressure $p = F_1 / A_1$, which generates a larger force on a piston of area A_2:

$$F_2 = p\,A_2 = F_1 \frac{A_2}{A_1}$$

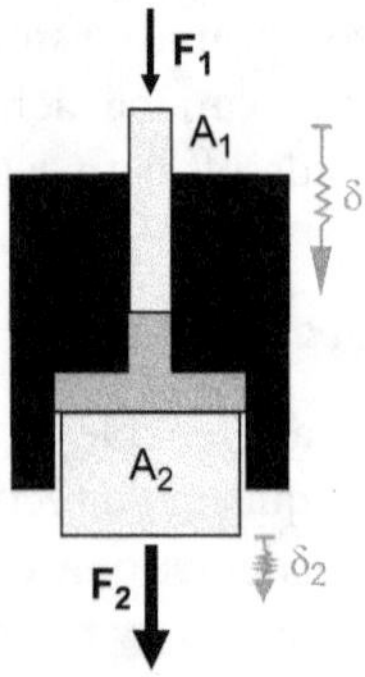

As always, force amplification is accompanied by a displacement trade-off. For an incompressible fluid, $\delta_1 A_1 = \delta_2 A_2$.

A particularly useful application arises when constant pressure is required despite volume changes due to thermal expansion, deformation or even leakage. A hydraulic press loaded with fixed weights provides a simple, robust solution: the weight sets the pressure, and the piston moves freely to accommodate volume changes; no sensors, electronics or active control are required.

Closing reflections

In today's research environment, experimentation is often associated with sophisticated instruments, sensitive sensors and massive data streams. This review of elemental machines is neither a historical exercise nor a retreat into simplicity for its own sake. Rather, it is an invitation to recognize the foundational ideas that underpin advanced systems and to anticipate trade-offs in the design of experimental devices.

The heart of experimentation is not rooted in complexity but in depth of thought, purposeful design and the transformation of meaningful questions into observable phenomena. *Research outcomes rely more on the clarity of the idea than on the perfection of the equipment. Simple devices make the purpose clearer, enabling insight, testing and iteration without wasting resources. Sophisticated instruments are valuable when the underlying phenomena are well understood and precise measurements are needed; but starting with elaborate and costly devices can enslave the researcher instead of helping to reveal the truth of an idea.*[5] In well-designed experiments, simplicity is not a constraint but a virtue. The most elegant experiments are lean, purposeful and grounded in fundamental principles.[6]

Exercises

1 Investigate the inner workings of experimental devices and sensors, from analog timers and pumps to the probe tips used in atomic force microscopes (AFM). Can you identify "classical simple machines" within these systems? Draw mechanically correct, clearly labeled sketches.
2 Describe and identify the applications of micrometric, tapered and ACME threads. Describe different types of screw heads and identify their uses. Compare locking nuts and jam nuts, and explain how they prevent loosening.
3 Take a 6 mm diameter screw, two nuts and two flat washers. Hold the screw by one nut in a vise. Then tighten the other nut until "something breaks". What broke? Observe carefully. What is the failure mode, for example, tensile fracture, thread stripping or washer deformation? Estimate the force you applied, compute the corresponding torque and verify it against the expected strength of the fastener.
4 Study an automotive differential, describe how it works and draw it by hand.

5 The Rubik's Cube was invented in 1974 by Hungarian sculptor and professor of architecture E. Rubik. How does it work? Describe the internal mechanism that allows independent face rotation while maintaining structural integrity.

Notes

1 Archimedes (287–212 BC), Greek mathematician, physicist, engineer and inventor.
2 See essay on *Linear Actuators, Motors and Pumps (Section 4).*
3 See essay on *Our Physical Universe: Laws and Principles (Section 1).*
4 This sketch is based on a picture sourced from the internet.
5 Paraphrased from K. Terzaghi, in R.E. Goodman (1999). *Karl Terzaghi: The Engineer as Artist.* American Society of Civil Engineers.
6 See essay on *Elegance in Experimental Research (Section 1).*

Essay 25

DEVICES

Design strategies enhanced by digital twins

The devil is in the details...![1]

Design is challenging, fun and fascinating! The conception of devices and test protocols is an inherently creative process, one that mirrors the stages of creative problem-solving: define, explore, refine and validate.[2] Some devices are designed for exploratory studies and are kept deliberately simple to focus on the purpose; others are engineered for extensive experimental programs, designed to automate tasks while minimizing bias from nuisance variables and accidental errors.[3] In both cases, the design of devices must be grounded in rigorous thinking within predefined practical constraints. The process is rarely straightforward; instead, it unfolds through iterative cycles of critical thinking, adaptation and gradual refinement. Some general guiding concepts and applied principles follow.

Part 1: the classical approach

Getting Started: Define the Problem

Begin with a well-defined question and a clear research strategy. What phenomenon are you trying to observe? What measurements are required? Next, focus on goals and constraints. Describe the key characteristics of the specimen, such as phase, composition, acidity or conductivity. Identify critical parameters, including the minimum specimen size needed for representativeness. Estimate the duration of each test across relevant variable ranges and the number of tests needed to explore the domain of governing variables.[2] Define the boundary conditions that must be met and the necessary

DOI: 10.4324/9781003727521-29

measurement resolution. Establish allowable fluctuations in environmental variables such as deformation, temperature, pressure, chemical composition and electrical potential, along with operational requirements. Take your time in this initial step; it will pay off.

Idea Generation

Generate multiple design concepts without investing too much time in any single idea. Aim to develop alternatives that differ from one another as much as possible. Draw inspiration from similar studies in other fields, and explore online resources for images, configurations and potential components.

Involve insightful individuals with complementary expertise to benefit from collective knowledge and cross-disciplinary insight.[4] However, at the initial stage, independent idea generation is particularly important to avoid anchoring effects.[5]

Sketch each concept and conduct thought experiments to anticipate its behavior and performance. Begin selecting potential materials. Consider both construction and operational demands, and adjust designs to accommodate manufacturing constraints; for example, traditional machining tends to favor axisymmetric geometries, while three-dimensional printing allows for far more complex shapes.

Continue developing the different design concepts, updating your evaluation criteria as needed. Analyze every option against the goals and constraints defined earlier. Identify their advantages and disadvantages and refine the designs to preserve strengths while addressing weaknesses. Gradually narrow the options to a smaller set of the most promising alternatives.

Design Refinement

Select materials that satisfy the mechanical, thermal, chemical, electrical and biological requirements, while also considering machinability. Identify and evaluate potential failure modes and their interactions. Use closed-form analytical expressions to estimate initial component dimensions. Incorporate off-the-shelf components, standard sizes and readily available materials to reduce costs and streamline construction. Verify the final design through numerical analysis to confirm its behavior under expected conditions.

Ensure that tolerances align with performance needs, machining methods and material properties. Remember that *tight tolerances → high cost.* Carefully consider whether a high level of precision is truly necessary; for example, do you really need tolerances so tight that machining must be done in strictly controlled ambient temperatures to avoid thermal effects? Could the requirements be relaxed to save time and cost without compromising function?

Refine your design iteratively and consider simplifications during optimization. Aim for a well-balanced system in which all potential failure modes exhibit similar safety factors or levels of risk. Incorporate redundancies, especially in critical systems, so that failure does not depend on a single component. Ensure compliance with applicable codes and institutional requirements: structural strength, allowable deformation, leakage, biosafety, etc.

Finally, review the design critically in light of the project's goals and constraints. Consider ease of construction, assembly and disassembly, safe and reliable operation, cost, maintenance and any secondary issues such as mechanical noise or compromised data quality. At this stage, attention to detail is essential; remember, *the devil is in the details*!

Prototype(s) and Final Design

Prepare detailed blueprints with multiple views and cross-sections before moving on to prototype fabrication. If you are building the device yourself, follow the classic advice: *measure twice, cut once*!

Assemble the device carefully and methodically. Verify safety by integrating appropriate monitoring systems and incorporating fail-safe protocols; for example, use liquids rather than gases during pressurization tests to reduce the risk of a catastrophic failure. Clearly define the allowable operating conditions, such as pressure and temperature limits.

Once assembled, operate the device and test its performance rigorously. Identify flaws or limitations to guide future refinements. Optimize by reassessing all potential sources of coupled excitations and confirming that tolerances remain appropriate.

Part 2: designing devices in the digital twin era

The design of devices must remain closely aligned with the systems and theories under investigation to ensure that the data collected meaningfully supports the overarching research goals. Devices and instrumentation should also be developed in coordination with the analytical methods planned for data analysis. Within this framework, numerical tools (particularly digital twins) are emerging as critical complements to physical experimentation in modern research.

A digital twin is a virtual replica of a physical system. It facilitates the optimization of geometry, material selection, and sensor range and resolution, while also enabling strategic sensor placement to maximize information gain. Additionally, it helps define boundary conditions and establish appropriate device tolerances. Just as advances in materials, shape optimization and digital simulation have transformed athletic performance (allowing running shoes, bicycles and swimsuits to achieve record-breaking efficiency),

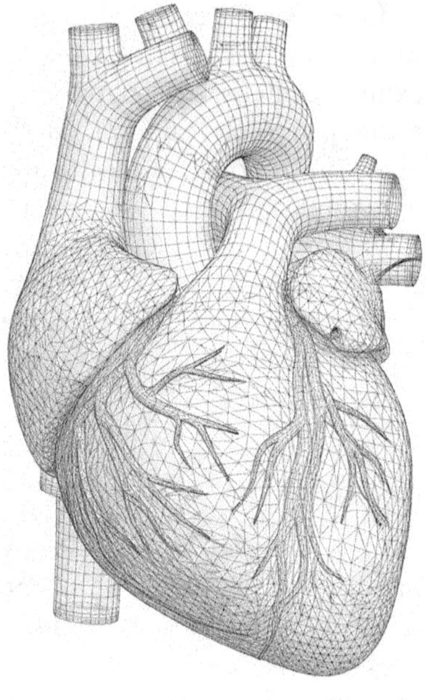

digital twins allow experimental devices to be designed and operated with enhanced capabilities.

Beyond design, digital twins support adaptive experimentation by providing real-time control under evolving boundary conditions. For example, consider a soil specimen representing the ground beneath a structure: running a physical test in parallel with a digital twin allows simulation of soil-structure interaction during an earthquake, with the twin dynamically updating the boundary conditions applied to the soil specimen.

When experiments involve coupled processes or highly nonlinear behavior, the physical laws embedded in the digital twin are essential for accurate data interpretation and integration. Take, for instance, an experimental study of an orthopedic implant subjected to hydro-thermo-chemo-mechanical coupling under complex boundary conditions: the digital twin provides a consistent framework to integrate diverse data streams.

Finally, once validated and calibrated, a digital twin can extend the reach of physical testing by simulating scenarios that are too costly, risky or impractical to reproduce in the lab. It also supports the extrapolation or upscaling of experimental results to real-world applications.

Clearly, the complementary use of digital twins offers significant advantages in cost, efficiency and insight far beyond what can be achieved through physical testing alone. While they do not replace hands-on experimentation, digital twins now stand as a powerful and indispensable tool in the modern experimentalist's toolbox.

Closing thoughts

Designing experimental devices is not merely about assembling components; it is about asking the right questions, defining meaningful constraints and translating abstract concepts into physical systems. Every sketch, material choice and instrumentation decision reflects your understanding of the phenomenon you seek to study. And in turn, thoughtful device design sharpens your thinking.

When approached with care and critical insight, the act of designing becomes an integral part of the reasoning process. Consider data analysis from the outset, not as an afterthought once measurements are in hand but as a guiding element in the experimental plan. In fact, complementing your experiments with analytical methods and numerical models, such as digital twins, will strengthen your research… you will fly much better with two wings!

Exercises

1 List seven different ways to measure a building's height. Aim to identify physically distinct methods.
2 Take pictures of seven different engineered parts or systems around you. Analyze/describe their geometric and mechanical design in detail (as well as functionality).
3 Design a device to retrieve a gold ring from a drainage pipe.
4 Without first reviewing existing solutions, design a device and deployment protocol for inserting a stent to restore blood flow through an artery. Use thought experiments to test your concepts, identify potential limitations, and refine your design. Then investigate the devices and procedures currently used in medicine. Compare your approach with existing solutions and discuss the underlying design principles.

Notes

1 Idiom – origin dates to the 19th century.
2 See essay on *Creativity and Creative Individuals (Section 2)*.
3 See essay on *Statistical Concepts for Experimental Design and Execution (Section 3)*.
4 See essay on *Research Teams: Creative Synergy (Section 2)*.
5 See essay on *Biases in Decision-Making (Section 2)*.

Essay 26

MATERIAL, FORM AND FUNCTION

Firmitas, utilitas, venustas
(Strength, usefulness, beauty)[1]

The subtle interplay between material, form and function is fundamental to the design of effective experimental devices. Nature offers countless examples in which geometry and material properties are intricately combined to achieve remarkable performance. Consider the eggshell: though seemingly delicate, its curved ellipsoidal shape allows it to withstand substantial pressure applied at its poles. Or consider the honeycomb, whose hexagonal geometry efficiently redistributes boundary forces, maximizes honey storage and minimizes the amount of beeswax required for construction.

The evolution of materials and forms

Throughout history, architectural forms have been shaped by the materials and tools available at the time. The ancient Egyptians constructed their massive pyramids by stacking large limestone blocks, relying on the natural strength of stone in compression. As engineering principles evolved in Egypt and Mesopotamia, new structural elements like columns and beams emerged to serve specialized functions. However, spanning large distances remained a fundamental challenge: stone is weak in tension, and increasing the thickness of a lintel merely shifts the failure mode from tensile cracking at the center to shear near the supports. The invention of the arch, later perfected and widely employed by the Romans, marked a critical breakthrough. By redirecting forces into compression along a curved path, arches enabled much larger

DOI: 10.4324/9781003727521-30

spans and paved the way for more complex architectural forms, including vaults and domes.

The introduction of new materials has continually expanded the possibilities for structural performance and design. Metals enabled tension-bearing structures such as cable-supported bridges. Steel also led to the use of I-shaped beams, where flanges resist tensile and compressive forces while a slender web carries shear with minimal weight. Reinforced concrete further optimized structural performance by combining the tensile strength of steel with the compressive capacity of concrete, as exemplified in T-shaped beams.

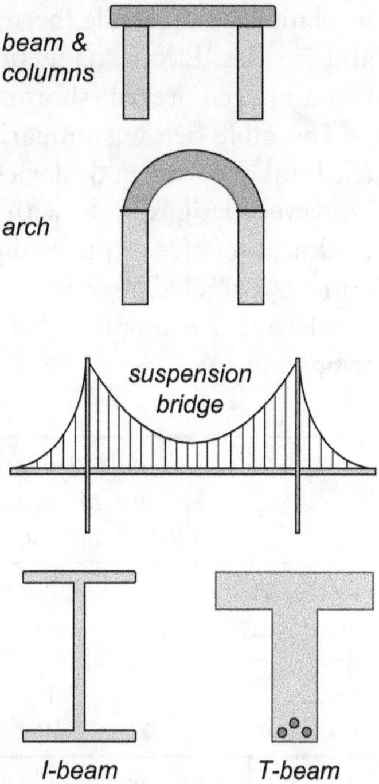

Material, form... and function!

At the heart of these examples lies the fundamental relationship between material, form and function, which carries through the engineered world. The wheel's circular shape enables seamless rotation and smooth motion with minimal resistance.[2] Similarly, today's flagpoles have a circular cross-section to withstand wind equally from all directions, while their conical shape effectively resists the linearly increasing bending moment imposed by the flag force.

Clearly, no shape can optimally address all design requirements. While spherical pressure vessels are preferred because they sustain uniform tension without flexural stress (in contrast, a prismatic pressure vessel would experience wall bending), most pressure vessels are cylindrical because they are easier to manufacture. Yet, cylindrical surfaces distort optical imaging, making square or flat geometries preferable, at least for low-pressure cells. Finally, feedthrough ports, holes and discontinuities create stress concentrations that negatively impact the mechanical integrity of all vessels.

Selecting materials

The design of experimental devices must consider material choices from the start to ensure their functionality, precision and reliability. Material mechanical stiffness and strength determine the device's compliance and integrity. Thermal conductivity must be consistent with required heat transport or insulation needs, while thermal expansion may cause unwanted deformation and stresses. Electromagnetic properties will affect shielding, prevent cross-talk and even prevent short circuits.

The table below summarizes selected properties of materials commonly used in experimental devices. Properties vary by orders of magnitude. However, designs scale with the governing material property, such as shell thickness relative to internal pressure, making the proper choice of materials critical. The challenge is further compounded when coupled processes are involved, for example, when tests experience large changes in pressure and temperature.

	Mass density [g/cm³]	*Young's Modulus [GPa]*	*Ultimate strength [MPa]*	*Thermal conduct [W/m·K]*	*Thermal expansion [10⁻⁶/°C]*	*Electrical conduct [S/m]*	*Magnetic perm [μ/μ₀]*
Steel SS-316♦	8	200	450	35	14	10^6	1
Aluminum♠	2.7	70	310	230	22	3×10^6	1
Glass	2.2	50	60♣	1	3.3	10^{-12}	1
Plastic	1	1	10–80	0.2	100	10^{-13}	1
Wood (pine)	0.5	10	40	0.2	4	10^{-12}	1

Nominal values from published handbooks and peer-reviewed studies; actual values may vary. Note: ♠ alloys. ♣ tensile. ♦ Super Duplex SS has twice the stiffness and strength and superior corrosion resistance than SS-316 but is partially ferromagnetic.

While most experimental devices use cost-effective, readily available materials that perform reliably under standard conditions, some applications require specialized materials with exceptional properties:

- *Borosilicate glass (Pyrex®).* Valued for its excellent chemical resistance, withstanding acids, bases and organic solvents (except hydrofluoric acid). Its low thermal expansion coefficient makes it highly resistant to thermal shock and high-temperature environments. Mechanically, it is stronger and more scratch-resistant than standard glass, though it remains brittle under impact. These properties make it ideal for analytical chemistry, petrochemical processes and reaction chambers.
- *Viton® rubber.* A high-performance fluoroelastomer with outstanding chemical resistance and thermal stability (–15°C to 205°C, up to 315°C intermittently). Low gas permeability and hydrophobic properties make it ideal for seals, gaskets and O-rings in extreme environments, though at a higher cost than conventional elastomers.
- *Sapphire.* It exhibits exceptional hardness, optical transparency and resistance to extreme temperatures and chemicals. Its high strength, scratch resistance and broad optical transmission range – from ultraviolet to

infrared – make it a preferred material for optical windows, laser applications and viewports in high-pressure setups.

- *X-ray transparent materials.* Used in applications where minimal X-ray interference is required. Beryllium provides high transparency and strength, but is costly and carcinogenic. Fused silica (quartz glass) combines chemical and thermal resistance with strength. Polyimide films (e.g., Kapton) are lightweight and flexible, ideal for low-pressure X-ray windows. PMMA (acrylic) is a cost-effective, versatile and transparent thermoplastic resin with moderate strength. On the other hand, PEEK is a high-performance opaque thermoplastic with superior mechanical, chemical and thermal properties.
- *Magnetically transparent materials.* These materials have a magnetic permeability close to free space $\mu \approx \mu_0$ and do not distort magnetic fields, so they allow for magnetic sensing. Examples include aluminum, copper and titanium or plastics such as polycarbonate for lightweight, non-reactive applications. For Nuclear Magnetic Resonance (NMR) characterization, the devices should neither interfere with magnetic fields nor with resonance signals. If the target is hydrogen protons, then avoid hydrogen atoms; in this case, suitable materials are borosilicate glass and quartz, polyethylene, polypropylene and Teflon® PTFE.

The following table shows their nominal properties for preliminary material selection considerations.

	Mass density [g/cm^3]	*Young's Modulus* [GPa]	*Ultimate strength* [MPa]	*Thermal conduct* [W/m·K]	*Thermal expansion* [10^{-6}/°C]	*Electrical conduct* [S/m]	*Magnetic perm* [μ/μ_0]
Pyrex®	2.2	64	70	1.2	3.3	10^{-13}	~1
Viton®	2	0.001	10	0.15	15	10^{-14}	~1
Sapphire	3.98	350	500	40	5.8	10^{-13}	~1
PMMA	1.2	2	70	0.20	70	10^{-14}	~1
PEEK	1.3	3	100	0.20	50	10^{-14}	~1
Teflon®	2.2	0.5	30	0.25	100	10^{-15}	~1

Nominal values from published handbooks and peer-reviewed studies; actual values may vary.

The soul of the story

Our experimental devices allow us to study phenomena under carefully controlled boundary conditions. Poor design can limit the scope of investigation, introduce observational bias and even create safety risks. In contrast, a well-designed device exhibits a distinctive beauty: a harmonious union of material, purpose and form. Ultimately, the choices we make in design reflect both our skill in crafting tools and our depth of understanding of the phenomena we aim to explore.

Exercises

1. Try to break an egg by pressing on its poles with your fingers. Can you do it? Explain why the egg is strong along this axis, considering its geometry and material properties.
2. Explore force redistribution in a honeycomb structure. If a unit force is applied at one corner of a hexagonal cell, how does the force propagate through the hexagonal network? Describe how this pattern of force redistribution contributes to the efficiency of honeycomb structures in load-bearing applications. Investigate common examples, from corrugated cardboard to energy-absorbing crash structures.

Notes

1. The "Vitruvian Triad" from Vitruvius (80BC–15BC), Roman architect and engineer.
2. See essay on *Enduring Simple Machines: From Archimedes to Galileo and Pascal (Section 4)*.

Essay 27

REACTION FRAMES AND VESSELS

Mechanical design principles

Ut tensio, sic vis
(As the extension, so the force)[1]

Test cells, pressure vessels and reaction frames are used across many disciplines to study processes under a wide range of pressure and temperature conditions. Designing these devices calls for a systematic approach that balances constraints and functionality. Begin by exploring multiple possibilities. Consider the operating conditions and potential test protocols. Some tests require the material to be chemically inert or corrosion-resistant, optically or magnetically transparent, or mechanically robust[2]... and sometimes all at once! Design soon becomes an art, where experience guides the prioritization of criteria. Here, we introduce fundamental concepts and analyses for the preliminary mechanical design of these devices. The expressions provided are standard and widely available in introductory mechanics and design textbooks; they are included for conceptual guidance and preliminary assessment.

Concepts from mechanics

Mechanical analysis builds on three sets of equations:

- *Equilibrium*. A system remains in equilibrium when the sum of forces $\sum F_i = 0$ and the sum of moments $\sum M_i = 0$ are zero. Uncertainties in material properties or loading conditions are accounted for by including a Factor of Safety (FoS) or, alternatively, by adopting a probabilistic analysis.

DOI: 10.4324/9781003727521-31

- *Compatibility of deformations*. Two connected points must undergo the same displacement to preserve continuity and maintain system integrity.
- *Constitutive equations*. There is a material-specific relationship between stress-and-strain $\sigma - \varepsilon$ and force-and-deformation $F - \delta$. For example, linear elastic materials follow Hooke's law, $\sigma = E\,\varepsilon$ where E is Young's modulus, or $F = k\,\delta$ where k is the spring constant. More advanced constitutive models capture nonlinear behavior, yielding, creep and anisotropy, among others.

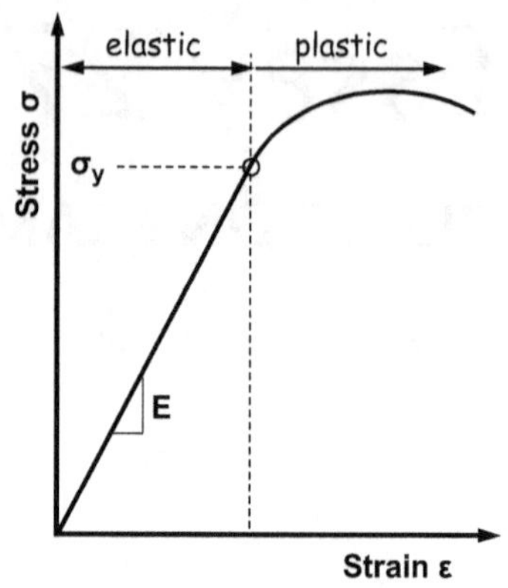

Equilibrium equations are sufficient to analyze isostatic systems, such as a beam on two supports, but are not sufficient for hyperstatic systems; for example, a cantilever beam with an additional deformable support at the free end. In this case, equilibrium alone is not enough; we must also invoke compatibility of deformations: the beam and spring displacements must be the same at their contact point, $d_{\text{beam}} = d_{\text{spring}}$. The resulting system of equations involves forces (equilibrium) and deformations (compatibility). To solve them together, we must convert forces-to-deformations or deformations-to-forces (constitutive equations).

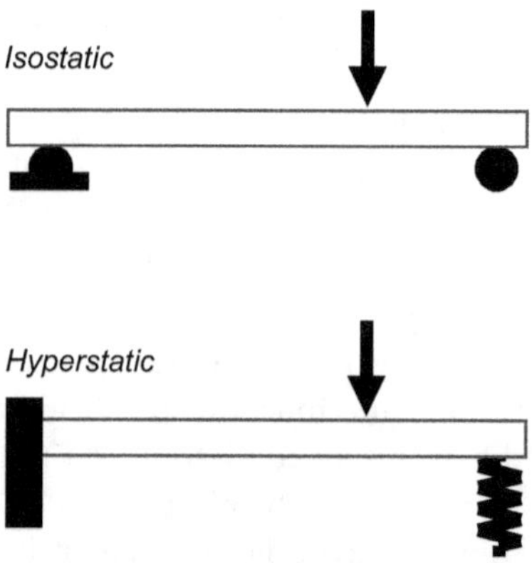

The coexistence of compatibility and equilibrium conditions in hyperstatic systems has important consequences. Consider a force N applied to a system of two soft springs and one rigid spring connected in parallel. Clearly, $N = F_1 + F_2 + F_3$ (equilibrium), but how much load does each spring carry? If crossbeams are rigid, all springs deform equally $\delta_1 = \delta_2 = \delta_3 = \delta$ (compatibility). Then, the force in each spring is $F_i = k_i\delta$ (constitutive equation), and we readily conclude that the stiffer central spring will carry a larger share of the load.[3] This observation can be generalized into a key guiding principle for mechanical design:

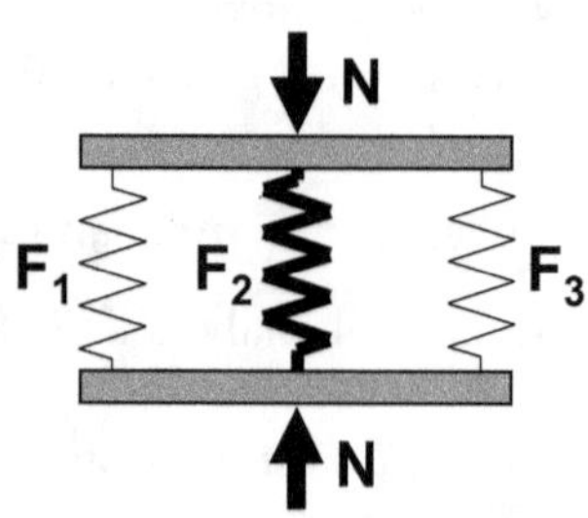

stiffer elements attract a greater share of the load (in hyperstatic systems)

Overall, mechanical design must limit *deformation* that could compromise performance or bias measurements. At the same time, it must ensure sufficient *strength* to minimize the probability of failure across all potential modes. Remember: *nature always finds the weakest link*! The design of reaction frames and pressure vessels exemplifies these critical requirements.

Reaction frame - Design principles

Consider the reaction frame sketched in the figure, typically used to apply either compressive or tensile loads to a specimen.

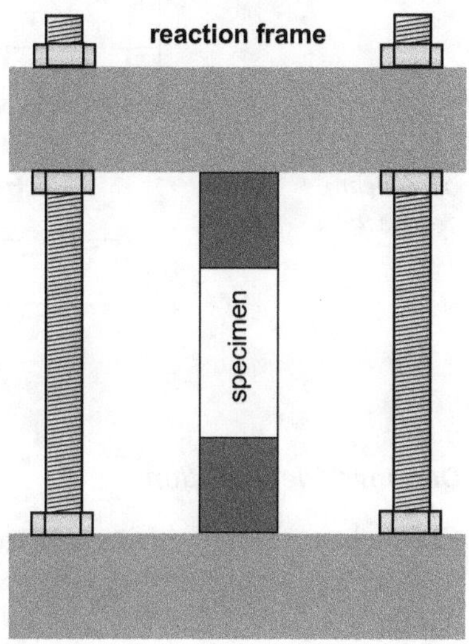

Design for strength – Redundancy

Evaluate every possible failure mode; then ensure that stresses remain below the material's yield stress σ_y everywhere, and apply an appropriate FoS to control the probability of failure for each potential mode.

Selecting the FoS is not trivial. It depends on acceptable risk: for standard designs, a FoS ≈ 1.5 may suffice; critical systems often require FoS > 2.5. Material variability and ductility also influence the choice: for the same application, mild steel or aluminum may justify FoS ≈ 1.5, while titanium alloys might demand FoS ≈ 3 due to lower ductility and higher variability. Always confirm applicable code requirements first.

To improve safety and reliability, *incorporate redundancy* by using components in parallel; for example, design a frame with four rods instead of two. The main failure modes and basic design formulas for frame design are outlined below.

rod in tension

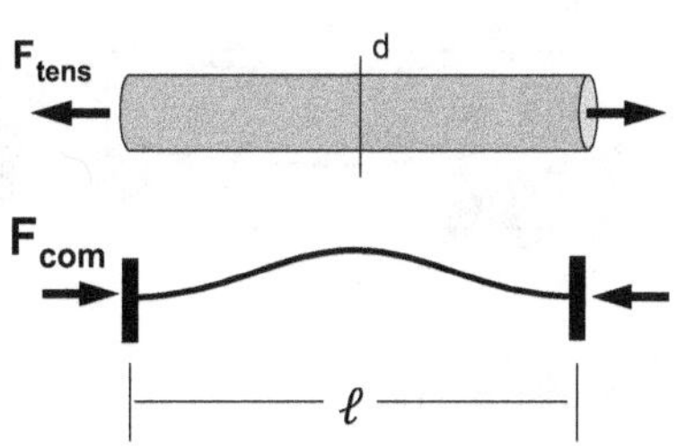

$$\sigma_{tens} = \frac{4}{\pi}\frac{F_{tens}}{d^2}$$

rod in compression – buckling

$$\sigma_{com} = \frac{4}{\pi}\frac{F_{com}}{d^2} \le 16\pi^2 E \frac{d^2}{\ell^2}$$

threads in shear

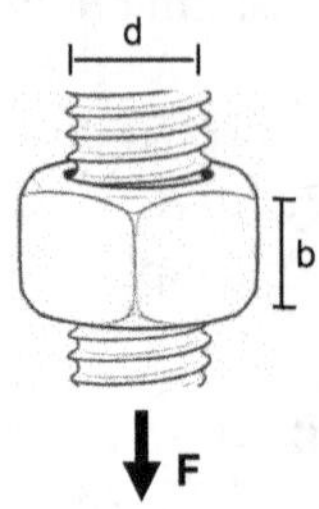

$$\tau_{\text{shear}} = \frac{F}{\pi d b}$$

cross beam in shear

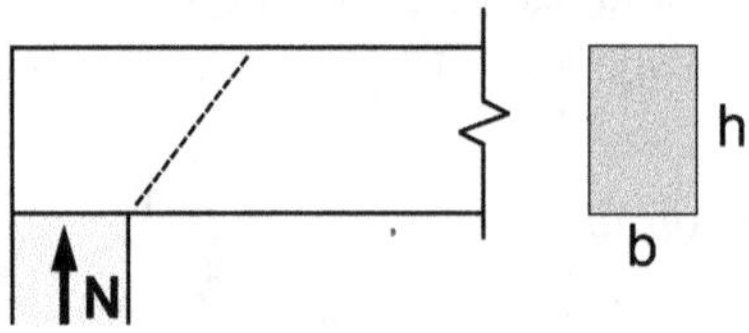

$$\tau_{\text{shear}} = \frac{3}{2}\frac{N}{bh}$$

cross beam deflection

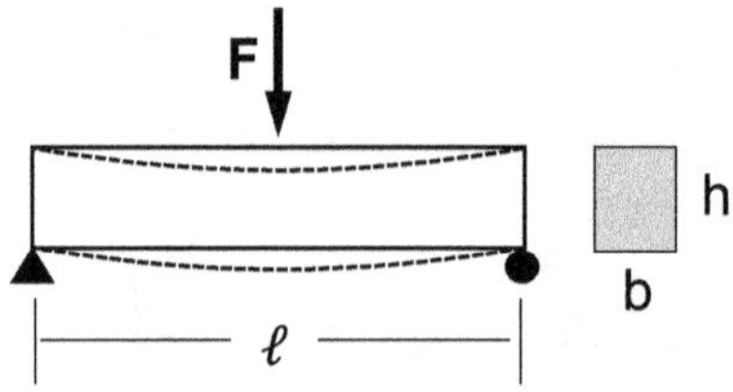

$$\sigma_{\text{bend}} = \frac{3}{2}\frac{F\ell}{bh^2}$$

Design for deformation

Even when a system meets strength requirements, deformations must remain within acceptable limits to preserve functionality and prevent permanent damage.

cylinders (rods and caps)

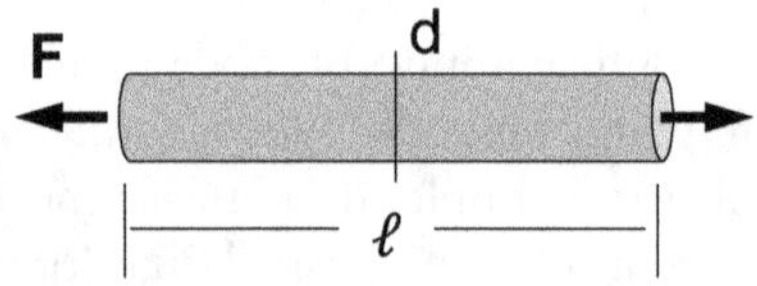

$$\delta_{\text{cyl}} = \frac{4}{\pi}\frac{F\ell}{d^2E}$$

cross beam deformation (simply supported beam)

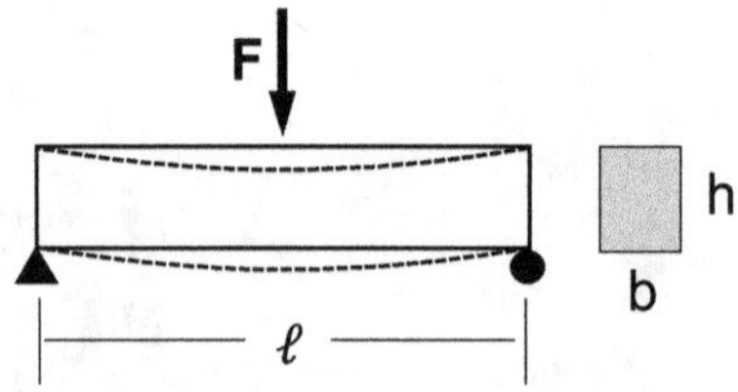

$$\delta_{\text{beam}} = \frac{1}{4}\frac{F\ell^3}{Ebh^3}$$

Acceptable limits must also account for the behavior of the material under study. For example, consider a material exhibiting post-peak relaxation over a deformation range denoted as δ_{relax}. The total frame deformation δ_{frame}

required to reach the peak load is the sum of all component deformations: $\delta_{frame} = \delta_{caps} + \delta_{cross\ beam} + \delta_{rods} + \delta_{bottom\ plate}$. If the frame is "soft" $\delta_{frame} >> \delta_{relax}$, the stored strain energy will trigger sudden specimen collapse (open circles in the sketch). To capture the material's post-peak strain-softening behavior (continuous line), the frame must be sufficiently rigid such that $\delta_{frame} << \delta_{relax}$.

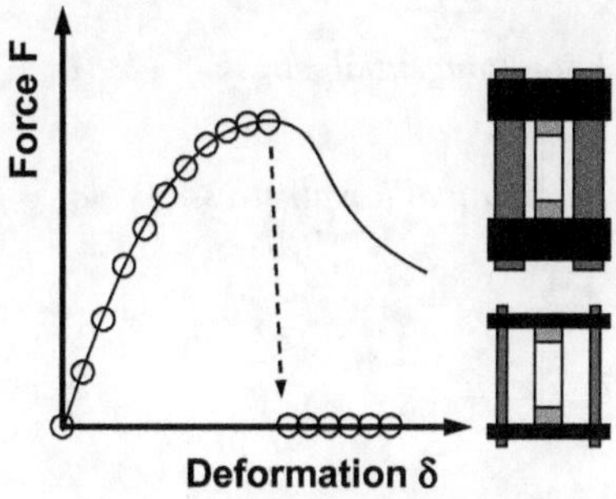

Advanced loading frames use servo-controllers to rapidly unload the specimen, effectively simulating a virtually infinite frame stiffness. These systems achieve precise mechanical control and prevent premature collapse by leveraging the much faster response of electronics compared to mechanical components.

Pressure vessel - Design principles[4]

Vessel walls can experience either compressive or tensile stress. *Compressive stress* arises when external pressure exceeds internal pressure, as in ultra-high vacuum vessels or deep-sea submersibles operating at depths of 10,000 meters, where external pressures can reach ~100 MPa. Conversely, *tensile stress* develops in vessels subjected to internal overpressure, from hyperbaric chambers used in medical recompression therapy (150–300 kPa) to autoclaves and high-pressure reactors for reaction kinetics and phase transition studies, which may reach several hundred MPa.

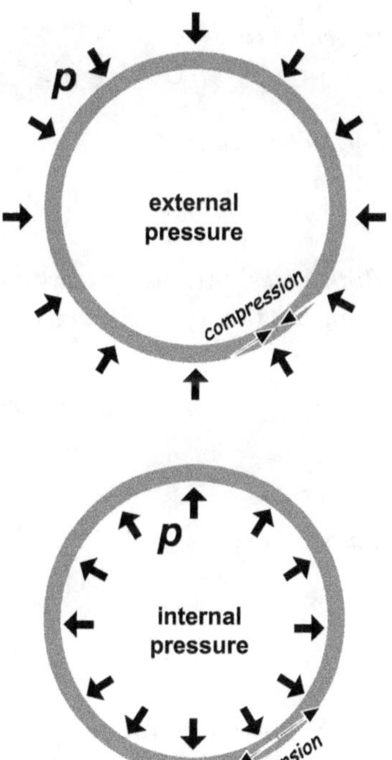

As the required wall thickness h increases relative to the vessel inner radius r_i, the stress distribution across the wall becomes non-uniform. Depending on the h / r_i ratio, either thin-wall or thick-wall analysis applies.

Thin-walled approximation ($h / r_i < 0.1$)

Hoop and axial stresses are uniform across the wall thickness, while the radial stress is negligible in comparison (Note: check wall buckling for thin-walled vessels under compression):

hoop (tangential) stress $\sigma_\theta = p\dfrac{r_i}{h}$

longitudinal (axial) stress $\sigma_\ell = p\dfrac{r_i}{2h}$

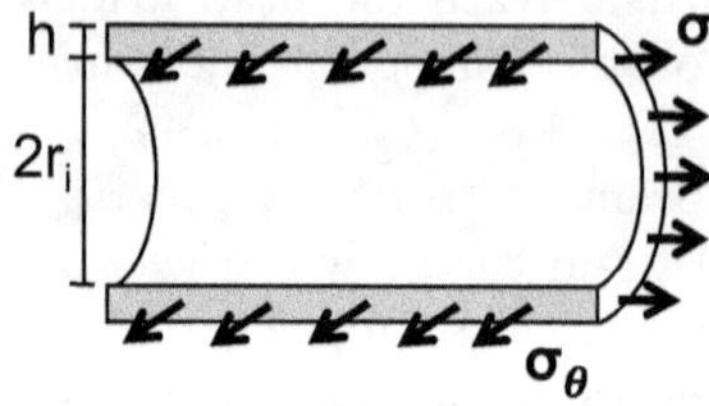

Thick-walled approximation $(h / a > 0.1)$

Stresses vary significantly across the wall thickness. This requires a more complex analysis that accounts for radial stress gradients:

radial stress $\sigma_r = p\dfrac{r_i^2}{r_e^2 - r_i^2}\left(1 - \dfrac{r_e^2}{r^2}\right)$

tangential stress $\sigma_\theta = p\dfrac{r_i^2}{r_e^2 - r_i^2}\left(1 + \dfrac{r_e^2}{r^2}\right)$

longitudinal stress $\sigma_\ell = p\dfrac{r_i^2}{r_e^2 - r_i^2}$

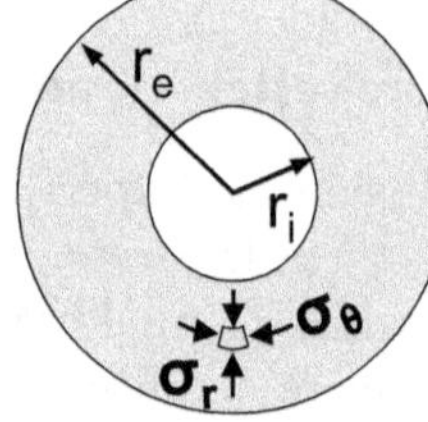

Both cases. Strains at working pressures (well below failure) are calculated using the theory of elasticity, taking Poisson's effect into account via Hooke's Law:

elastic strains $\varepsilon_x = \dfrac{1}{E}\left[\sigma_x - \nu\left(\sigma_y + \sigma_z\right)\right]$

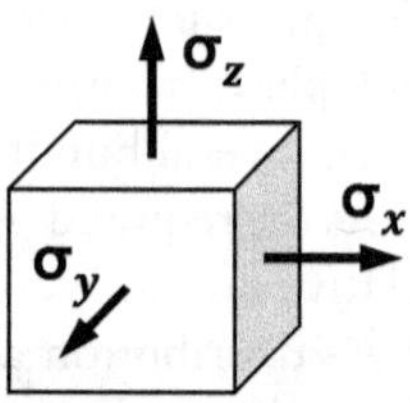

where Poisson's ratio ranges from $\nu = 0.28$–0.33 for most metals.

The vessel design includes peripheral components such as tubes and valves. Anticipate operational scenarios and prioritize safety at every stage. For instance, is there a risk of unintentionally cross-connecting high- and low-pressure lines? If so, incorporate safeguards and components such as pressure relief valves to prevent potential hazards.

Pressure containment

O-rings are essential components in pressure systems. Compressed between two mating surfaces, they deform to fill the gap and form a reliable seal. Their effectiveness rests on an elemental observation: even high pressures can be contained with minimal force if the sealing gap is small. Yet, O-rings must also withstand challenging chemical and thermal conditions, making both material selection and geometric design critical for performance.

Material selection

It must consider pressure, temperature, mechanical displacement and chemical exposure, as each factor can cause swelling, embrittlement or wear. The Space Shuttle *Challenger* disaster stands as a sobering reminder of the consequences of inadequate O-ring performance under critical environmental conditions.[5] Consider:

Nitrile NBR	oils and fuels; operational range –40°C to 120°C
Viton® FKM	chemically aggressive environments (fuels, acids); –20°C to 200°C
Polyurethane	high mechanical strength and abrasion resistance for dynamic seals
Silicone	extreme temperatures (–60°C to 230°C); avoid in high-wear or petroleum fluids
EPDM	compatible with water, steam and some chemicals; not for oils
PTFE	exceptional chemical resistance; limited elasticity
Metal-based	extreme conditions (materials include nickel, copper, aluminum, indium and others)

Based on published standards and manufacturer datasheets, final material selection must be verified for application-specific conditions.

As an example, consider O-rings for CO_2 vessels. Material choice must account for operating pressure, temperature and the resulting CO_2 phase (gas, supercritical, liquid or solid). NBR and neoprene should be avoided because they tend to swell, form blisters and ultimately fail from explosive decompression. More suitable options include Viton® FKM or PTFE-encapsulated O-rings.

Groove geometry and design

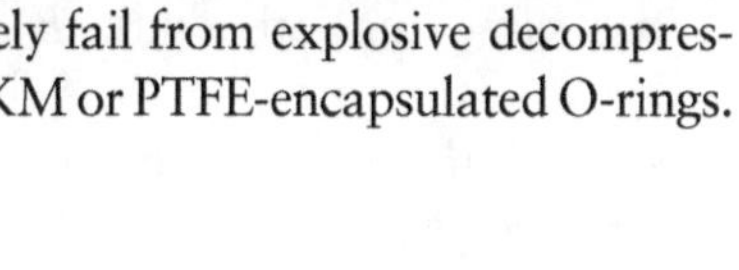

Sealing performance depends critically on the geometry of the housing groove and the clearance between mating components. O-ring deformation must be carefully controlled: insufficient compression may result in leakage, while excessive compression can cause premature wear or permanent deformation. A common design target is 10%–30% squeeze, with groove depth 75–85% of the O-ring's cross-sectional diameter and groove width ~1.5 times the diameter. Surface finish also matters; a typical roughness average is Ra ≈ 1μm.

In dynamic applications, such as pistons or rotating shafts, lower compression is preferred to reduce friction and wear. Vacuum-compatible sealing grease can be used, provided it is chemically compatible with both the O-ring and the working fluid. Clearance gaps should allow assembly but must prevent O-ring extrusion; at pressures above ~10 MPa, backup rings are often employed to support the O-ring. Finally, always ensure that grooves and mating surfaces are clean and free of debris before assembly to maintain sealing integrity.

Thermomechanical coupling

When components are constrained, even small temperature changes can generate significant internal forces that compromise both structural integrity and measurement accuracy. Consider two rods with different Young's modulus E, cross-sectional area A and thermal expansion coefficient α, fixed between two rigid walls and constrained against lateral deformation. The internal response to a temperature change ΔT depends on their configuration:

- *Parallel configuration.* The rods act independently and mobilize different forces. The thermal expansion that each rod would undergo freely $\alpha \, \Delta T \, \ell$ is opposed by the elastic deformation caused by the emerging reaction force from the rigid constraint $F\ell/(EA)$, resulting in net zero deformation. Therefore, the internal force in each rod is $F_i = \alpha_i \, A_i \, E_i \, \Delta T$.

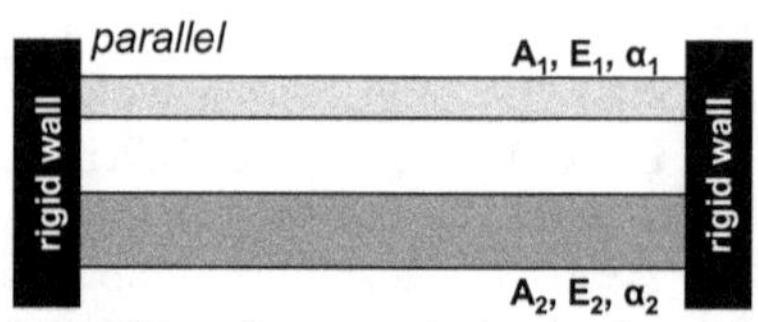

- *Series configuration.* Force equilibrium requires that the axial force be the same in both rods, $F_1 = F_2 = F$. Because the total length is fixed, the sum of the free thermal expansions must be exactly balanced by the elastic contractions induced by the emerging internal force F:

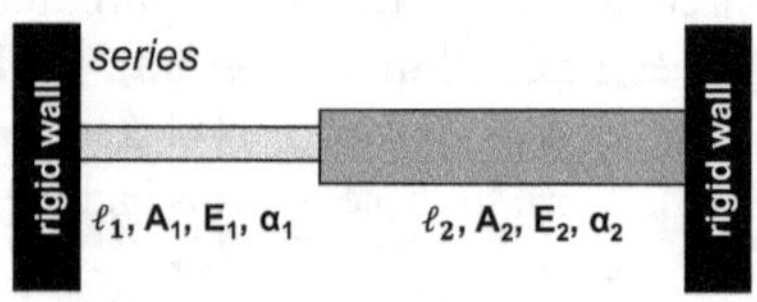

$$F = \frac{\alpha_1 \ell_1 + \alpha_2 \ell_2}{\dfrac{\ell_1}{A_1 E_2} + \dfrac{\ell_2}{A_1 E_2}} \Delta T$$

Thermal effects are particularly critical in multi-material assemblies, where differences in thermal expansion can cause stress concentration, misalignment or even mechanical failure.

Repetitive loads? Fatigue!

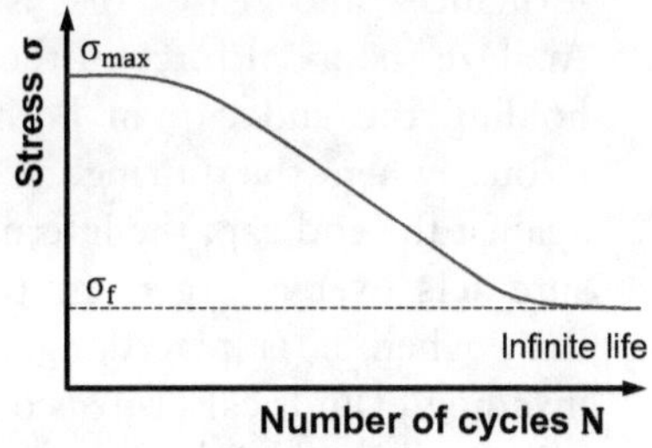

The maximum weight we can lift in a single effort represents our strength; we must lighten the load to lift it multiple times. Mechanical components behave similarly: under repeated thermomechanical cycling, they must carry less than their single-load capacity to avoid fatigue failure.

The relation between the maximum stress σ and the number of cycles to failure N is represented by the σ –N Wöhler curve. If the applied stress σ exceeds the material's fatigue limit σ_f, damage accumulates and failure can occur even after a relatively modest number of cycles. Designs for fatigue-sensitive applications should therefore adopt lower allowable stresses.

So many more thoughts...

Always consider the potential interactions between mechanical, thermal, chemical, hydraulic and biological processes; for example, corrosion may emerge as the limiting factor in some systems. Ensure compatibility between components and operating conditions during assembly, operation and maintenance. And above all, prioritize safety.

The most effective experimental setups are purposeful, balanced and elegant.[6] You will know you have a great design "*not when there is nothing left to add, but when there is nothing left to take away*".[7] Remain mindful. Remain deliberate. Remain thoughtFULL!

Exercises

1 Design a cylindrical confining cell capable of subjecting a 10 mm diameter specimen to an internal pressure of 1 MPa. The cell must be X-ray transparent to allow in situ monitoring using high-resolution X-ray micro-computed tomography (micro-CT). Select materials that meet both transparency and mechanical strength requirements. Identify all potential failure modes (e.g., yielding, buckling, fracture, fatigue and leakage). Design the cell to achieve a balanced safety margin, aiming for similar factors of safety across all failure modes.
2 The Bridgman seal can withstand very high pressures (up to 40,000 MPa).[8] How does it work?
3 Evaluate the optimal location for an O-ring in a cylindrical pressure vessel: should it be placed against the end cap or against the invading cylindrical protrusion of the cap? Discuss the advantages and disadvantages of each

configuration in terms of sealing performance and ease of assembly. Analyze the axial force in the screws holding the end cap in both cases. (Note: when the O-ring is located against the end cap, the internal pressure acts over a larger effective area than when it is placed against the invading cylindrical protrusion of the cap, resulting in a higher axial force pushing the cap outward.)

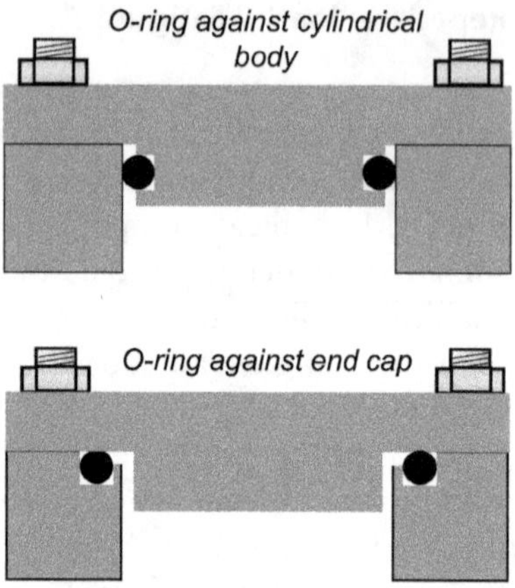

Notes

1 R Hooke (1635–1703), English polymath. Hooke's law implies a linear relationship between force and deformation $F = k \cdot \delta$
2 See essay on *Material, Form and Function (Section 4).*
3 Combining the three sets of equations, $\delta = N / (k_1 + k_2 + k_3)$.
4 Refer to the codes published by the American Society of Mechanical Engineers ASME, Section VIII: Division 1 for pressures below 20 MPa, and Division 2 for higher-pressure applications.
5 The Space Shuttle Challenger disaster (January 28, 1986) was attributed to O-ring embrittlement caused by record-low temperatures on the day of the launch.
6 See essay on *Elegance in Experimental Research (Section 1).*
7 A. de Saint-Exupéry (1900–1944), French writer, poet, journalist and aviator.
8 P.W. Bridgman (1882–1961) received a Nobel Prize in Physics for his work on high pressure.

Essay 28

LINEAR ACTUATORS, MOTORS AND PUMPS

Motion is the essence of life[1]

In experimental research, motion is rarely an end in itself; rather, it is a means to control, stimulate or observe a process. Whether shifting a stage, rotating a specimen in a tomographer or displacing fluids in a microfluidic chip, we rely on linear actuators, motors and pumps to convert various forms of energy into precise, controlled motion. Their design exemplifies creative engineering grounded in fundamental physical principles. Let's explore some of these devices to uncover the core concepts that drive their operation and inspire alternative designs.

Linear actuators

Linear actuators convert various forms of energy[2] into linear displacement. Common examples include:

- *Mechanical*: Transform mechanical energy into motion using mechanisms like levers, screws and gears.[3] Reliable and well-suited for low-speed, low-frequency applications.
- *Hydraulic*: Convert pressurized liquid flow into mechanical force. The near-incompressibility of liquids enables high forces and smooth, controllable motion, ideal for large mechanical systems, including position control.
- *Pneumatic*: Use the energy in compressed gas to produce motion, providing clean actuation. They offer greater compliance than hydraulic systems, but with lower force capacity and stiffness. Favored in automation, lab

DOI: 10.4324/9781003727521-32

equipment and lightweight robotics, where simplicity and speed outweigh precision.

- *Electromagnetic*: Convert electrical energy into motion through magnetic fields. In solenoids, a current through a coil creates a magnetic field that shifts a central ferromagnetic plunger. They deliver considerable force and speed, from small-scale instruments to industrial automation.
- *Electrostatic*: Generate motion by attraction or repulsion between electrically charged elements. Though forces are small, they are effective at micrometer scales, such as in microelectromechanical systems (MEMS). Their compatibility with semiconductor fabrication allows seamless integration into microchips.
- *Thermal*: Produce motion through temperature-induced deformation. Basic actuators rely on linear thermal expansion; bimorphs consist of two layers with different coefficients of thermal expansion, so one expands more than the other when heated, causing bending. More advanced systems use shape memory alloys like Nitinol. Compact and forceful for their size, thermal actuators are suited for low-frequency applications such as thermal triggers, thermal MEMS micro-actuators and classical thermostats.

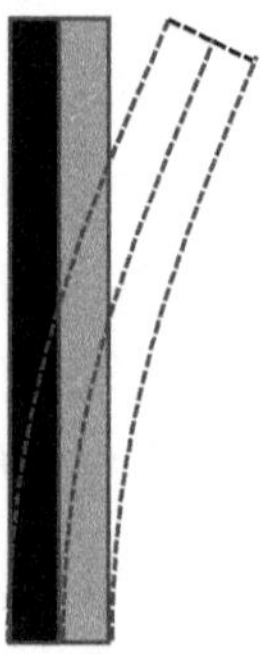

- *Piezoelectric*: Use materials that deform when subjected to an electric field. These actuators offer micron-to-nanometer-scale displacement control with high speed and repeatability, making them essential in optics, scanning probe microscopy and microfluidics. Piezoelectric bimorphs or "bender elements" bond two piezoelectric layers together to produce large tip displacements compared to the small strain within each layer. Their motion is precise but limited in range.
- *Magnetostrictive*: Rely on materials that change shape when exposed to magnetic fields. Although the deformation is small, it is rapid and forceful, enabling use in high-frequency applications such as sonar, vibration control and precision positioning.
- *Chemical and biochemical*: Use materials that deform through chemical reactions, such as swelling from pH changes, phase transitions (e.g., ice-water) or variations in suction (as in hydrogels). These actuators may incorporate biological materials or microorganisms like motile bacteria. Operating without electrical input, they can be highly sensitive to chemical environments and even light. Applications range from drug delivery and nanomotors to soft robotics and tissue engineering.

Motors

Motors can deliver precise torque, speed and position control, making them ubiquitous across scales: from nanomotors in microfluidic devices to industrial servos in robotic arms and testing machines. Motors convert electrical

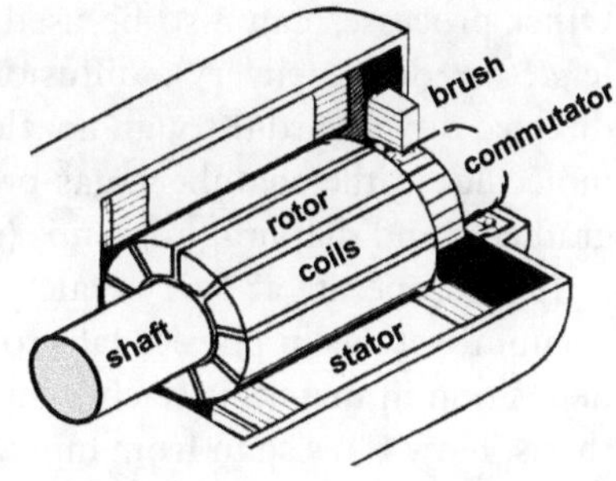

energy into rotational motion. At their core, they operate through the interaction of magnetic fields and electric currents: when current flows through a coil in the presence of a magnetic field, a force is generated, producing torque and causing rotation. The *stator* is the stationary part of the motor that creates the magnetic field (either a permanent magnet or coils), while the *rotor* consists of coils mounted on the rotating *shaft*. Although there are several motor types, they all follow the same underlying physical principle[4]:

- *DC motors*. The commutator shifts the current in the rotor to keep the motor turning. Speed is proportional to the applied voltage and torque to the current. Easy to control, DC motors are common in low-voltage lab systems, portable devices and quick prototypes.
- *Brushless motors*. Use electronic commutation to energize coils in the stator in sequence, eliminating brush contacts. This improves durability and performance in high-reliability systems.
- *AC motors*. The alternating current in the stator creates a rotating magnetic field that drives the motor. They run efficiently and continuously, making them common in industrial environments, especially where speed regulation is less critical.
- *Stepper motors*. Move in discrete steps, typically ~1°, by controlling the coils that are energized in the stator. This allows open-loop position control without feedback, ideal for applications requiring repeatable positioning.

Pumps

Pumps convert mechanical energy into fluid motion, enabling circulation, injection and pressure control. Among the earliest designs, Archimedes' screw (3rd century BC) used a rotating helical shaft to lift water: an elegant design still found in many modern pumps. Centuries later J. Bramah's hydraulic press[3] (18th century) laid the foundation for modern hydraulic systems. Two principles govern the operation of mechanical pumps:

- *Positive displacement pumps* trap and move fixed volumes of fluid per cycle. They provide precise flow control and operate reliably under varying pressures. Rotary, reciprocating and peristaltic pumps fall into this category. Our own heart is a reciprocating positive displacement pump.
- *Rotodynamic pumps* impart velocity to the fluid and convert it into pressure. Their continuous operation suits high-throughput, low-viscosity systems.

Other processes can also be used to move fluids, including osmotic pumps (e.g., for drug delivery), diffusion pumps (vapor generated in a boiler is directed downward through nozzles, forming high-speed jets that capture gas molecules in the chamber), gas-pressure-driven flow using imposed pressure gradients and diaphragm pumps (a form of positive displacement).

Pumps operate at every scale and under every condition: from ultra-high vacuum systems in physics labs to perfusion control in bioreactors, thin-film deposition in materials science and flow chemistry setups in automated synthesis. Flow rates span from hundreds of liters per second in industrial applications and deep wells to picoliters per second in microfluidics. Design must balance pressure, flow rate and the fluid's physical and chemical properties: viscosity, temperature and corrosiveness all influence pump selection and material compatibility.

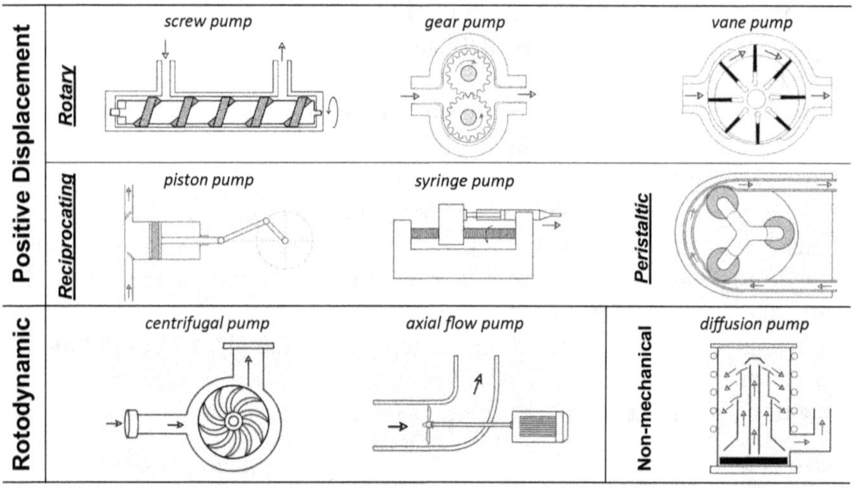

Note: The sketches depict standard pump concepts and are based on images found in technical literature. Analyze them by "following the fluid".

Output control

Inherent limitations in actuation systems (such as thermal effects and gear backlash) and nonlinearities and progressive aging (wear, demagnetization, depolarization and fatigue) affect accuracy over time. High-quality components and good engineering can minimize these effects; still, high-precision actuation and automation require sensing and feedback. In *closed-loop control*, the output is continuously monitored and fed back into the system to automatically adjust the input and correct deviations to maintain the desired performance. Examples include servo motors and pumps operating under flow or pressure control.

Closing thoughts

In experimental design, carefully selected actuators, motors and pumps can greatly enhance a system's precision, repeatability and responsiveness. Their integration, however, demands thoughtful consideration of factors such as power, inertia, coupling to other components and control strategies (open-loop vs. closed-loop).

All devices face an inherent trade-off between range and resolution: maximum displacement versus precision, torque versus angular resolution or volume versus flow rate. Effective experimental design means matching the right devices to the specific purpose. Ultimately, the question is not just, "can it move?" but rather, "can it move as much as needed, as fast as needed, and with the precision required?"

Exercises

1 Carefully disassemble a spare linear actuator, motor or pump, removing all non-glued/welded components. Document key stages with photographs, practice proper tools/techniques, analyze mechanical and electronic operation, and reassemble, ensuring correct alignment and torque. Prepare a pictorial guide and have a colleague follow it. Observe the device's engineering design.
2 Study an Ariete (hydraulic ram) pump, which lifts water without electricity/fuel using the water hammer effect. Analyze how the valve harnesses pressure surges to move fluid, and make a detailed sketch.
3 Design a high-pressure, high-precision syringe pump (piston, sliding seal, reaction frame, screw-driven actuator) capable of injecting acidic fluids at 1 μL/s, 20 MPa, 20 mL total capacity. Include detailed mechanical and sealing considerations.
4 Design a hydrogel-based actuator responsive to chemical changes. Explore geometries for significant motion and assess repeatability under cyclic or varying conditions (Note: hydrogels are commonly found in diapers; try experimenting with them).

Notes

1 Motion lay at the heart of early Greek philosophy and can be traced to Heraclitus (535 BC–475 BC).
2 See essay on *Our Physical Universe: Laws and Principles (Section 1).*
3 See essay on *Enduring Simple Machines: From Archimedes to Galileo and Pascal (Section 4).*
4 This sketch is based on an image sourced from the internet.

Essay 29

IMPEDANCE

Opposing energy flow

Energy is the currency of the universe, never lost, only transformed[1]

Impedance z measures a system's resistance to energy flow. It is defined as the ratio of an input (such as force or voltage) to an output (such as deformation or current). In quasi-static or steady-state conditions, this ratio is a real number (e.g., stiffness or resistance). However, when the input varies over time, the output typically lags, introducing a phase shift. In such cases, impedance must capture both the amplitude ratio and the phase difference between input and output. This is conveniently handled using complex numbers, hence the term *complex impedance*... even though there is nothing inherently "complex" about the concept!

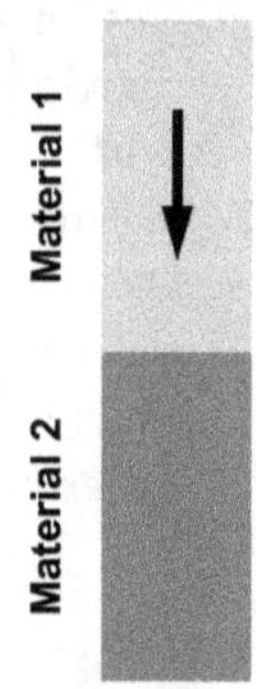

Now consider two adjacent materials or systems subjected to a time-varying input approaching their interface. How much of the input energy will be transmitted through? Will there be a phase shift? The answers depend on their *relative impedance* z_2 / z_1. Full transmission occurs when there is no impedance mismatch, that is, when $z_2 / z_1 = 1.0$.

DOI: 10.4324/9781003727521-33

Types of impedance

Mechanical impedance plays a critical role in fields such as vibration analysis and robotic actuation. In plane wave propagation, a medium's *acoustic impedance* z_a is the product of its mass density ρ and its wave propagation speed v. At the interface between two media, their relative impedance determines how much of the wave energy is transmitted and how much is reflected. This concept underpins a range of applications, including ultrasound imaging, non-destructive testing and noise control.

$$z_a = \rho v$$

Electromagnetic impedance governs how electric and magnetic fields interact with materials; in the case of light, it is referred to as *optical impedance*. A medium's electromagnetic impedance z_e is defined by the ratio of its magnetic permeability μ to its electric permittivity ε. Electromagnetic impedance underpins systems such as electromagnetic shielding (to reflect or absorb waves), cavity resonators, high-frequency circuits, wireless communications and fiber optics. In wires, transmission lines and antennas, impedance matching ensures efficient power transfer, minimizes reflections and maximizes fidelity, especially at frequencies above 1MHz.

$$z_e = \sqrt{\frac{\mu}{\varepsilon}}$$

Thermal impedance describes a material's resistance to heat flow. It is inversely proportional to its thermal conductivity k_T. Under steady-state conditions, the characteristic length L^* is the material's thickness. However, for thermal oscillations with period T, it is given by $L^* = \sqrt{T D_T / 2\pi}$ where D_T is the thermal diffusion coefficient. Thermal impedance plays a critical role in temperature management and stability in applications such as electronic systems and CPUs, batteries, wearables, calorimetry and sensors of all kinds.

$$z_T = \frac{L^*}{k_T}$$

The equations above assume one-dimensional energy flow and define a *specific* impedance that is independent of cross-sectional area. In practice, however, geometry plays a central role in impedance matching (hence, the importance of antenna shape).

Contact impedance can also arise when different transport mechanisms meet. For example, there is *thermal contact impedance* between a metal (where heat is carried by diffusing energetic electrons) and a dielectric (where heat is carried by phonons), and *electrical contact impedance* between a metal electrode (electron conduction) and a salt solution (ion conduction). *Coupling agents* are often used to improve impedance matching: gels and elastomers for mechanical waves, matching networks and transformers in electrical circuits, refractive index-matching liquids for optical testing, and thermal grease or phase-change materials for heat transfer.

In closing...

Neglecting impedance can undermine experimental outcomes: it causes energy losses, lowers signal-to-noise ratios, introduces signal artifacts, generates unwanted vibrations, contributes to overheating and may even lead to premature system failure. In multifunctional systems such as optoelectronic sensors or thermoacoustic devices, impedance mismatches can severely limit overall performance. Clearly, impedance is not merely a physical concept, nor is impedance matching just a design detail... it is central to the design of experimental devices. In fact, it often becomes a critical design constraint and a performance bottleneck.

Exercises

1 Feel it ... explore it ... analyze it:

  a Explore the effect of shoe inserts (e.g., gel insoles). Compare walking or running with and without gel insoles. Pay attention to differences in comfort, impact and fatigue.
  b Try hitting a punching bag with and without boxing gloves (*Caution*: start with soft punches to avoid injury!). Notice the change in impact felt in your hand, wrist, arm and shoulder.
  c View insoles and boxing gloves as mechanical impedance matchers. Design an experiment to measure their efficiency. Consider variables like material type and thickness.
  d Develop an analytical formulation (Hint: start from conservation of momentum).

2 Study P-wave transducers for air and water applications. Identify examples of water-matched transducers and investigate their design.
3 Explore coupling agents used in medical ultrasound. What properties make them effective?
4 Study how a quarter-wavelength matching layer (thickness $h = \lambda / 4$) reduces wave reflection at an interface. Follow the analytical formulation in detail to determine the required impedance of the matching layer and explore applications such as anti-reflective coatings, ultrasound probes and waveguides.

Note

1 Paraphrased from the Law of Conservation of Energy, formalized by German physicist H. von Helmholtz (1821–1894).

Essay 30

BOUNDARIES

Framing experiments in space and time

Circumstances shape every action; context cannot be ignored[1]

Experimental results are often shaped more by boundary conditions than by the intrinsic behavior of the system itself. While we readily acknowledge their critical role in solving differential equations or running numerical simulations, we are often less rigorous in specifying and controlling boundary conditions in experimental studies. Why is that? Boundary conditions are no less important in physical experiments: they are essential for producing valid, interpretable and reproducible results. Moreover, maintaining similarity at the boundaries is essential when using small-scale physical models to study larger prototypes.[2]

Admittedly, boundaries in experimental settings are difficult to assess and manage: complex setups, sensor placement and measurement limitations can all interfere. But these challenges should not become excuses. Early and deliberate consideration of boundary conditions significantly improves both data quality and interpretability. Although the discussion that follows draws mainly from engineering and the physical sciences, the underlying concepts are broadly applicable and equally relevant to experimental work in other fields, from the social sciences to economics.

Inspired by analysis

Once again, consider differential equations.[3] Ordinary differential equations involve a single independent variable (typically time) and its derivatives; hence, they require temporal boundary conditions. In contrast, partial differential equations describe systems that vary in both space and time and therefore require both spatial and temporal boundary conditions.

DOI: 10.4324/9781003727521-34

The main types of boundary conditions in analytical models either specify the value of the function on the boundary (*Dirichlet conditions*, e.g., surface temperature of a solid object), define the normal derivative of the function (*Neumann conditions*, e.g., heat flux through a surface), or impose a combination of the function's value and its derivative (*Robin conditions*, e.g., both temperature and heat flux at a boundary). Each type yields a distinct system response and is equally applicable in experimental contexts:

- Mechanics: stress-controlled, deformation-controlled or stiffness-controlled.
- Fluid dynamics: pressure or flow rate control.
- Thermodynamics: constant-pressure (isobaric) or constant-volume (isochoric).
- Thermal systems: adiabatic (no heat flux) or isothermal (constant temperature).
- Electrical systems: constant-voltage, constant-current and matched-impedance.
- Chemical systems: pervious vs. impervious (permeability), reactive vs. non-reactive (reactivity), constant mass closed-system vs. open-system.

Experimental implementation

Boundary conditions impose constraints on a system's behavior at the spatial limits of its domain. The concept extends equally to time-dependent processes, where initial and final conditions define the system's temporal evolution.

Spatial boundaries

Some experiments are conducted under closed boundary conditions, where all spatial boundaries are the same. However, experiments involving the flow of energy, mass or charge typically engage both longitudinal and transverse boundaries:

- *Longitudinal boundaries* align with the primary flow direction. These are often imposed but not always well-controlled. For instance, contact impedance can arise at the interface between the experimental system and the measuring device.[4] Alternatively, the device itself may fail to maintain target conditions, such as injecting water quickly enough to sustain a flow-controlled test, delivering constant electrical current or unloading a piston fast enough to achieve deformation-controlled loading in brittle materials.
- *Transverse boundaries* act perpendicular to the primary direction of the process. Although often assumed to play a passive role, in practice they

rarely do. In mechanical testing, for example, transverse boundary conditions may be designed as zero-lateral strain, plane-strain, plane-stress or constant-stress conditions; however, the actual stress or strain fields across the boundary are often non-uniform, and unintended shear stresses and strains can develop. Similarly, in fluid dynamics, transverse boundaries may be assumed to permit slip, but in reality, they often impose no-slip conditions or become preferential flow pathways. Processes observed along transverse boundaries are therefore particularly prone to misinterpretation due to unintended boundary effects.

Temporal boundaries

Time-dependent processes cannot be assessed or modeled without precise knowledge of the system's starting point and final state:

- *Initial conditions.* Experiments must begin from a clearly defined and well-characterized state. This requires careful specification of the system's initial condition at the outset and precise measurement of all relevant parameters.
- *Terminal conditions.* The state of the system at the conclusion of the experiment is often most informative. Terminal conditions help anchor the solution and validate models, support parameter estimation through back-propagation, and are essential for assessing energy or mass loss as well as any irreversible changes in state. They may also reveal internal processes or localized transformations that are not apparent from boundary measurements alone. Accordingly, careful disassembly accompanied by concurrent measurements is a critical final step in the experimental sequence.

Practical considerations

What strategies can experimentalists employ to minimize boundary effects and maximize the quality and interpretability of their data?[5]

Improve boundary conditions through creative (and rigorous) experimental design

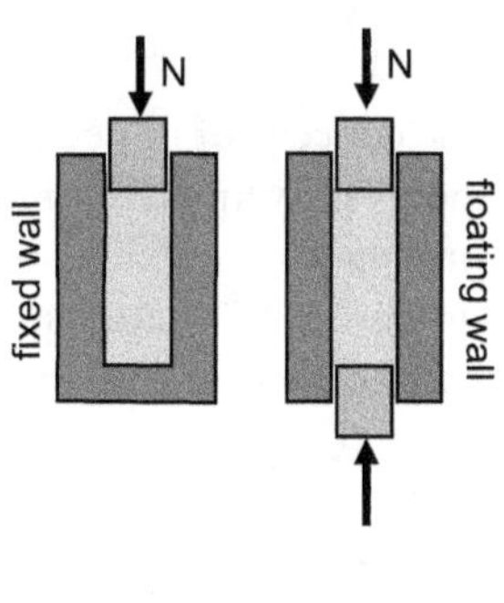

In mechanical systems, friction-reducing layers can significantly limit unwanted shear stresses at boundaries (some lubricants have friction coefficients as low as 0.05). Alternative geometric configurations also help: for instance, replacing a fixed wall with a floating wall further reduces frictional drag and enhances boundary control.

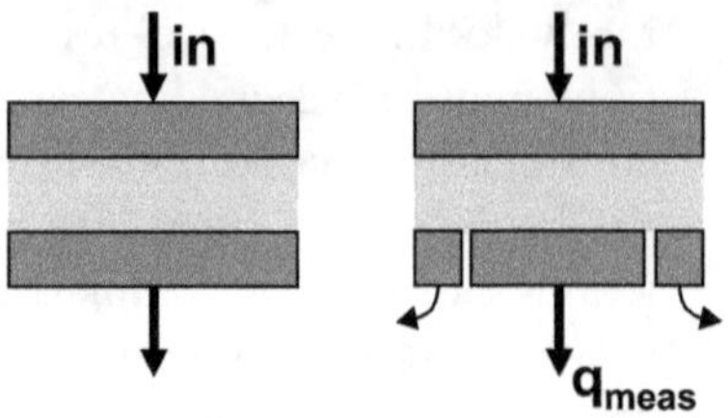

In fluid flow experiments, preferential flow along transverse boundaries can be eliminated using a double-ring permeameter, which discards any flow that reaches the outer annulus of the exit port; permeability is then calculated from the measured flow rate q_{meas} and the reduced cross-sectional area. Similarly, in electrical conductivity measurements, guarded-electrode setups effectively reduce surface leakage currents (Note: in both cases, the internal flow field is not strictly one-dimensional).

Increase the specimen size to extend the distance between the measurement zone and the boundaries

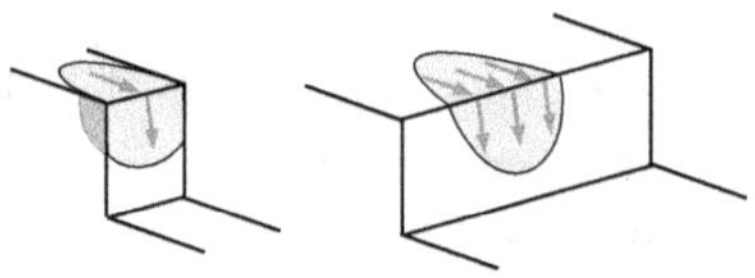

Some experiments are designed to collect data away from boundaries, assuming that "expected conditions have been rectified" within the measurement zone. In mechanics, this concept reflects Saint-Venant's Principle[6]: localized effects diminish with distance from the load application point.

But how far should boundaries be from the measurement zone? Rules of thumb provide guidance: a slenderness ratio of 8:1 is common for electrical conductivity tests, 5:1 for tensile strength tests and 2:1 for compression tests (a compromise between minimizing boundary effects and avoiding buckling).

However, geometry selection is not always straightforward and can vary even for the same problem under different conditions. Consider a bookshelf: when tightly packed, inserting one more book generates a pressure field that propagates far beyond the insertion point; yet, when the books are loosely arranged, the disturbance affects only the immediate neighbors, allowing boundaries to be placed much closer.

Match impedance[4]

The concept of impedance mismatch at boundaries is especially important in energy flow and dynamic conditions like wave propagation. Matching impedance helps minimize reflections, effectively creating anechoic boundaries. However, be cautious: material properties may change during the experiment, potentially causing impedance mismatches over time.

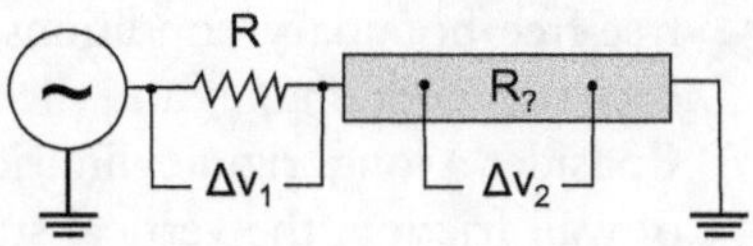

Thoughtful experimental design and precise instrumentation can also mitigate contact issues at boundaries. Consider electrical resistance R: the current i remains constant through the specimen due to charge conservation and can be reliably determined by measuring the voltage drop Δv_1 across a known resistor, $i = \Delta v_1 / R$. The voltage drop Δv_2 measured within the specimen, away from the boundaries, then provides the unknown resistance between electrodes: $R_? = \Delta v_2 / i = R\left(\Delta v_2 / \Delta v_1\right)$. This four-point method is not limited to electrical resistance; it can also be applied effectively to thermal and hydraulic conductivity measurements.

Modeling the measurement with its boundaries – Digital twin[7]

Because ideal boundary conditions are virtually impossible to impose, experimentalists must involve numerical simulations to capture the actual experimental boundary conditions (whether directly measured or hypothesized within a probabilistic framework). Advances in computing power have made this approach feasible, and it can be further enhanced by full-field monitoring technologies such as tomography. Together, these tools enable experimental designs that yield reliable physical insights from complex, real-world systems.

The punchline

Poor boundary control can introduce unintended biases, lead to misinterpretation of results, and cause discrepancies with theoretical predictions or simulations. Boundary conditions must therefore be carefully considered from the very start of experimental design. Otherwise, uncertainties will propagate through the data and undermine reliable interpretation. Trying to understand data without clear boundary details is like trying to follow a comedian in a foreign country: one may catch the words but miss the punchline.

Exercises

1 Compute the pressure that builds up in a liquid (such as water) under constant-volume (isochoric) conditions due to a 1°C increase in temperature. Discuss the potential implications for experimental design, particularly in sealed or rigid containers.

2 Study the resonant frequencies and vibration modes of a cantilever beam fixed at one end and free at the other. Compare them to a beam with

free-free boundary conditions. Analyze how boundary conditions influence the mode shapes and the natural frequencies.

3 Consider a long, rigid cylindrical silo filled with dry grains. In the absence of wall friction, the vertical stress $\sigma_z = \gamma z$ increases linearly with depth z, where γ is the unit weight of the granular medium. However, wall friction transfers part of the vertical load to the silo walls. Can you estimate the actual vertical stress distribution with depth? This is the classical silo problem described by the "Janssen solution".

4 Imagine the one-dimensional propagation of a freezing front under a constant surface temperature boundary condition. Sketch the expected temperature profile $T(z)$ at various times. Next, consider a spherical gelatin ball exposed to the same constant temperature at its surface. Analyze and compare the solution to the "Stefan problem" in both 1D and spherical geometries. Can you anticipate any emergent phenomena?

5 Explore the load-deformation response of cylinders made of the same elasto-plastic material but with different slenderness ratios. In each case, the cylinders are glued to rigid upper and lower caps and loaded under deviatoric stress until failure. Construct a digital twin to simulate the measured response (*forward problem*). Then, use the digital twin to interpret experimental data and attempt to extract the intrinsic material properties independent of boundary effects (*inverse problem*).

6 Investigate the use of electrorheological fluids to modulate boundary conditions in a system that requires tunable interfaces.

Notes

1 Reflects contextualism; inspired by J. Dewey (1859–1952), American psychologist and philosopher.

2 See essay on *Similarity and Scaling Laws (Section 3).*

3 See essay on *Differential equations: Processes in Mathematical Form (Unstated Assumptions) (Section 5).*

4 See essay on *Impedance: Opposing Energy Flow (Section 4).*

5 Note the parallel between experimental and numerical strategies for mitigating boundary effects, such as placing boundaries far from the region of interest or incorporating slip boundary elements to reduce friction.

6 J.C.B. de Saint-Venant (1797–1886), French mechanician and mathematician.

7 See essay on *Disruptive Technologies: Rethinking Methodologies and Shifting Paradigms (Section 3).*

Essay 31

NOISE … WHAT A PAIN!

Quiet the noise to reveal the truth[1]

Noise does not hide; it is always present in plain view, interwoven with every measurement we make. It blurs signals and obscures the very phenomena we aim to observe. In an age where sensors are inexpensive, data is abundant and computer algorithms are often misleadingly effective, it is tempting to rely on post-processing to clean up measurements. However, preventing noise through careful experimental design remains the most effective and reliable approach to ensuring data quality.

Noise reduction begins with experiment design

The fundamental strategy for noise control is simple: improve the experiment at its core. The following guidelines may appear to be listed chronologically, yet their order actually reflects their priority for obtaining reliable measurements. As you will see, the most effective steps for reducing noise occur well before data gathering.

Design the experiment to minimize noise and maximize signal-to-noise ratio (SNR). (1) Select variable ranges where the phenomenon produces a clear response distinguishable from background fluctuations: chemo-mechanical coupling is stronger in colloids than in millimetric particles. (2) Use geometry to amplify the signal: a longer rod yields greater accuracy when measuring thermal expansion than a shorter one. (3) Control the environment by employing both passive and active noise-control methods, as discussed later.

DOI: 10.4324/9781003727521-35

Choose the right transducers at the best location. (1) The transducer's full-scale range should be 1.5-to-2 times the maximum expected input to prevent saturation and maximize resolution (remember: there is an inherent trade-off between range and resolution).[2] (2) Select transducers based on functionality: avoid accelerometers when monitoring low-frequency oscillations, as high-frequency noise will dominate the output.[3] (3) Every sensing system has a response time: choose sensors with a response time much shorter than the characteristic time of the process you want to capture. (4) Be cautious with sensors that are sensitive to more than one concurrent process: fiber optic strain sensors and strain gauges can detect both strain and thermal changes. (5) Optimize sensor location to capture strong-amplitude signals, minimize boundary effects and extract the most information about the system.

Select peripheral electronics that meet the experimental requirements (power, frequency, resolution, etc.). Place emphasis on quality: good electronics yield better data and speed up experimental development. For example, low-cost multichannel A/D boards may advertise high resolution (in bits) and fast sampling rates; however, they often suffer from poor channel isolation, allowing crosstalk between channels.

Match impedances.[4] Impedance mismatch at active and passive boundaries, in sensor-medium coupling and along peripheral electronics reduces the signal amplitude, introduces reflections and degrades the SNR.

Carefully select or design chambers and devices.[5] Consider factors such as stiffness or compliance, resolution, and feedback control capabilities. Choose materials that ensure thermal and chemical stability under the operating conditions.

Calibrate devices. Regardless of the equipment (whether it's a standard pH meter or a sophisticated spectrometer), regular maintenance and calibration are essential to ensure accuracy and reliability over time.

Input and output. (1) Explore different input excitations: in linear systems, a sinusoidal input generates a sinusoidal output at the same frequency, which is easy to identify even when it is buried in noise; this strategy is central to spectral techniques. (2) Consider the sampling strategy based on the process under study: if it produces an exponential signal over time, collecting data on a logarithmic time scale may be more effective than sampling at constant intervals. (Note: avoid this approach if post-processing will involve standard frequency-domain analysis).

Use signal stacking to increase SNR.[6] Stacking-and-averaging is a highly effective method for noise reduction. Repeat the same excitation multiple times and sum the resulting output signals: random noise with a zero mean gradually cancels out, while the consistent signal becomes stronger. This method requires a repetitive source and a stable trigger, and it assumes the system remains time-invariant throughout the measurements.

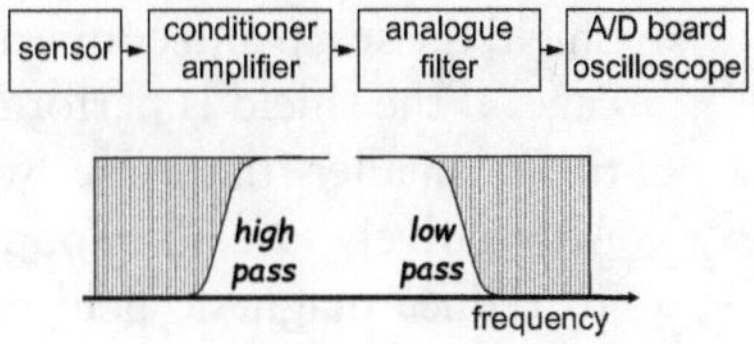

Include analog inline filters before A/D conversion but proceed with caution. Filtering will permanently affect the recorded signal. Set the high-pass filter to the lowest practical frequency and the low-pass filter to the highest, ensuring that the desired signal passes through with minimal distortion.

Signal processing.[5] Signal processing can empower the expert but mislead the amateur. It is essential to have a clear understanding of the mathematical operations and algorithms involved. Whether using a moving kernel in the time domain or windows in the frequency domain, signal processing can significantly alter the signal. Overlay the filtered signal with the original: the "cleaned" signal should retain the main features of the original without significant distortion.

Statistical techniques. Imagine gathering results from an infinite number of test replicas. The mean of all the outcomes would be the true value, and the standard deviation would reflect measurement variability. In reality, we can only conduct a limited number of tests: the probability that the computed mean is sufficiently close to the true value increases with the number of tests but also depends on the standard deviation. This observation underlies statistical methods in experimental design and data analysis.

Types of noise → Passive and active noise control

Effective noise control starts with a clear understanding of its sources, whether environmental, protocol-dependent or intrinsic to the equipment (e.g., mechanical, thermal and electromagnetic noise from peripheral electronics and actuators). Below, we outline various types of noise and discuss both passive and active methods for controlling them.

Electromagnetic noise

Electromagnetic noise is among the most pervasive sources of interference in laboratories, often originating from the very electronics used in the experiment.[7] The primary strategy is grounding, but it can be tricky. First, note that "ground" does not necessarily mean $V = 0$; in most laboratories, it is contaminated with a wide range of signals. Therefore, it is best to think of the ground as a common reference potential for all devices and measurements.

- *Passive systems.*
    - Connect ALL devices in your experiment to a common power strip to establish the same reference potential. Do not share the power strip with other ongoing tests.

  - Shield the setup and connect the shield to the ground to create a Faraday cage. If the shield is perforated for visibility, keep hole sizes at least 10 times smaller than the wavelength of the unwanted interference. Alternatively, use aluminum foil.
  - To reduce magnetic field interference, conduct tests inside enclosures made of high magnetic permeability materials (e.g., Mu-metal).
  - Cables act as antennas. To minimize interference: (1) keep them as short as possible; (2) twist and coil them to cancel induced currents, effective up to ~20 MHz; (3) use coaxial cables with shields connected to a common ground, effective from Hz to GHz.
  - For low-noise operation, battery-powered devices in a floating-ground configuration may be preferable.
  - Design measurement systems for noise cancellation; for instance, differential amplifiers help suppress common-mode noise.
  - Include inline analog filters (with caution, as discussed above).
- *Active systems.* While active electromagnetic noise cancellation systems exist, their use remains limited in experimental setups.

Mechanical noise

Vibrations can originate from moving parts within the experimental device (such as actuators) or from external sources (such as people, ventilation systems or nearby traffic).

- *Passive systems.* The classic strategy for vibration control is to isolate the base. Floating optical tables and hanging tables use a heavy mass m and low-stiffness supports k to create bases with a low resonant frequency $\omega = \sqrt{(k / m)}$. Geometric filters, such as sound diffusers or trenches, reduce surface wave propagation, while internal reflections are minimized using absorbing materials. Common components for acoustic noise control include sound-absorbing panels, foam wedges, heavy drapes and bass traps.[8]
- *Active systems.* Active mechanical noise control exploits the higher speed of electromagnetic transmission compared to mechanical wave propagation. These systems sense incoming vibrations and generate counteracting forces or sound waves with reversed polarity, either globally or at specific detection points. Noise-canceling headphones are a well-known example of this approach.

Thermal noise

A general rule of thumb is that most laboratory devices are excellent thermometers! Ambient temperature fluctuations (hourly, daily, weekly or seasonal) and heat generation by experimental devices can significantly impact a wide range of processes, from cell partitioning in biological experiments to high-pressure generation in liquids contained within rigid cells. Furthermore, thermally induced charge vibrations generate thermal noise in electronics (Johnson-Nyquist white noise).

- *Passive systems*. (1) Reduce heat transfer by blocking conduction and convection using low-thermal-conductivity materials such as foams, ceramics or vacuum layers and by limiting radiation with reflective coatings. (2) Increase thermal inertia by using large masses with high heat capacity, such as copper or aluminum blocks; water baths can also be used but should be covered to prevent evaporation. (3) Pay particular attention to conductive heat transfer along cables, tubing, fluid injection systems and mounts. Materials like PEEK, Delrin or aerogels can help minimize thermal transfer. (4) Avoid exposure to direct sunlight, airflows or HVAC vents. (5) Minimize internal heat sources, such as CPUs and lights.
- *Active systems*. Closed-loop control systems use temperature sensors in combination with heating or cooling elements (such as fluid circulation or Peltier coolers/heaters) to maintain stable thermal conditions (often within ±0.1°C).

Chemical noise

The laboratory environment, material contamination or unexpected reactions can all introduce chemical noise that alters experimental results.

- Check reagent purity to minimize interferences.
- Control reaction conditions, including pressure, temperature and pH.
- Moisture control is particularly important in processes sensitive to suction, adsorption and water-mediated reactions. Reliable passive methods can provide long-term stability: reduce humidity with dehumidifiers, which condense moisture on cold surfaces, or with desiccators filled with drying agents. Conversely, introduce moisture with humidifiers or by bubbling air through aqueous solutions, or use salt baths to maintain stable humidity. Commercial temperature and relative humidity controllers can further enhance precision.
- Use chambers filled with inert gases to create oxygen-free environments and prevent oxidation-related issues, and clean rooms to minimize particle interference from dust and aerosols.
- Chemical noise can also arise from interactions between transducers and the medium itself, such as electrode corrosion during long-term electrical conductivity measurements.

Biological noise

Bio-contamination can undermine not only biological studies but also a wide range of physical and chemical experiments. To minimize the risk, follow standardized protocols and enforce strict contamination control measures. This includes maintaining sterile preparation areas, adhering to cleaning routines and ensuring aseptic conditions for tools, surfaces and specimens. Additionally, limit sample handling time and exposure to environmental factors. Finally, make effective use of statistical approaches to account for any variability.

To emphasize once again

Noise is a challenge best addressed early in the experimental design process. Passive methods often provide the most effective and reliable solutions. Conduct preliminary tests to refine the design and ensure the collection of high-quality data. Remove outliers before applying signal processing or statistical analysis, but proceed with caution, as not every unexpected fluctuation is noise; some may reflect the true complexity of the system's response. Post-processing techniques, while potentially powerful, can also be misleading.

Finally, believe it or not, noise can sometimes be useful. If you are curious, explore the fascinating phenomenon of stochastic resonance, where noise can actually enhance the detection of weak signals.

Exercises

1 The holes in a microwave window allow visible light to pass through (optical wavelengths) while preventing microwaves from escaping the resonant cavity where food is heated, acting like a Faraday cage. Measure the diameter of the holes and compare it to the wavelengths of both light and microwaves. Discuss your findings. (Hints: speed of light $c_0 = 3\times10^8$ m/s and the operating frequency of microwave ovens is 2.45 GHz).

2 Design a vibration isolation table that filters frequencies above 10 Hz. The table, which should have a high mass, can either rest on soft rubber pads or be suspended from the ceiling using elastic cords. Analyze it as a single-degree-of-freedom system, calculate and plot its frequency response, and determine the resonant frequency. Vary the table mass and the stiffness of the support system to meet design criteria. Finally, compare your design with commercially available vibration isolation tables.

3 Daily temperature fluctuations can exceed 2°C in research laboratories, with even larger variations during weekends (HVAC systems, sunlight, building occupancy and equipment use). Design a passive thermal

isolation system to maintain a stable temperature within ±0.1°C for a 1-liter pressure vessel. Consider the mounts, piping and fluid injection/extraction systems. Focus on the details to ensure effective temperature regulation.

Notes

1 Inspired by spiritual and mindfulness writings.
2 Essay on *Sensing the Experiment: Scientific Sensors (Section 4).*
3 Assume a sinusoidal displacement $x = A\sin(\omega t)$; the acceleration is $a = d^2x / dt^2 = -\omega^2 A\sin(\omega t)$, thus, high-frequency noise is amplified by ω^2.
4 See essay on *Impedance: Opposing Flow (Section 4).*
5 See essay on *Reaction Frames and Vessels: Mechanical Design Principles (Section 4).*
6 See essay on *Signal Processing and System Analysis (Section 5).*
7 See essay on *Electronics: Fundamentals and Applications (Section 4).*
8 When combined, these components can create "anechoic chambers", which approximate free-field acoustic conditions by minimizing reflections from surrounding surfaces.

Essay 32

SENSING THE WORLD AROUND US

Our natural sensors

All our knowledge begins with senses[1]

Human senses are exceptional. Shaped by evolution, each sense serves a unique and indispensable function, enabling us to adapt and interact with our environment. None is redundant; none is an evolutionary accident. Together, they form a multi-sensory, multi-physics interface that detects, compares and interprets the diverse signals encountered as we navigate the world around us.

Similarly, we instrument experiments to monitor the phenomena under investigation.[2] This parallel invites us to look to our own bodies for inspiration, recognizing human perception as a remarkable natural prototype.

Our senses

The following highlights explore the core functions of our senses[3]:

- *Sight*: Human vision is limited to a narrow band of the electromagnetic spectrum, with wavelengths between approximately 380 and 750 nanometers. It balances high-resolution color detection in daylight (enabled by roughly 5 million cone cells in the retina) with high-sensitivity grayscale vision at night (facilitated by ~110 million rod cells, about 1,000 times more sensitive than cones). Collectively, these photoreceptors allow the human eye to operate across ten orders of magnitude in luminance.
- *Hearing*: The auditory system detects pressure waves within the frequency range of 20 Hz to 20 kHz. While we cannot perceive the ultrasound of bats or the infrasound of elephants, we are highly sensitive to the tonal and temporal features of speech and to acoustic cues in our environment.

DOI: 10.4324/9781003727521-36

- *Taste*: The human tongue contains more than 5,000 taste buds, each with approximately 100 chemoreceptors. Taste perception is limited to five modalities (sweet, salty, sour, bitter and umami[4]) but plays a vital role in risk detection (such as identifying toxins) and guiding food choices.
- *Smell*: The olfactory system contains roughly 10 million receptors high in the nasal cavity (dogs, by comparison, have 20-to-30 times more). These receptors can distinguish a wide variety of scents, with some substances detectable at concentrations as low as parts per trillion.
- *Touch*: Tactile sensing relies on more than 100,000 mechanoreceptors, each specialized to detect pressure, vibration, texture and stretch. Calibrated more for protection than precision, this system allows us to detect harmful extremes without overwhelming the brain with minor fluctuations.

Unique characteristics

Salient features of our sensory systems can inspire the design of experimental instrumentation:

- *High receptor density with overlapping sensitivity*: For example, rod and cone cells in the retina operate across complementary lighting conditions, extending the eye's dynamic range.
- *Multi-physics capability*: Human senses capture electromagnetic, acoustic, chemical (both airborne and dissolved) and mechanical stimuli, providing a comprehensive environmental profile.
- *Trade-off between range and resolution*: Like all sensors, human senses face an inherent trade-off between range and resolution. Note that sensory resolution is not the same but varies across modalities[5]: vision ~9 bits, hearing ~20 bits, taste ~4 bits, smell ~40 bits and touch 5 to 10 bits.
- *Response time compatibility*: Sensory response times align with our body's time scales. We can perceive flashes as short as 10 to 20 milliseconds and detect certain smells in about one millisecond. For comparison, physical reactions such as a finger snap occur in as little as 10 milliseconds.
- *Sensory adaptation and self-calibration*: Sensitivity adapts to sustained stimuli, for example, temperature perception shifts after holding a cold object. Furthermore, we quickly adapt to darker and quieter surroundings when leaving a noisy party late at night.
- *Spatial distribution and directional sensitivity*: Receptor placement is optimized for function. For instance, fingertips contain a dense array of mechanoreceptors. Spatial distribution also enables depth perception through stereoscopic vision reconstructed from slight differences between images captured by each eye.[6]
- *Protective feedback loops*: We rapidly withdraw our hand upon touching a hot object. These built-in mechanisms safeguard both the sensors... and us.

Perception = sense + processor

Perception is not just passive sensing; it requires active processing. The brain can isolate a single conversation in a noisy room (the "cocktail party effect") and localize a calling friend using subtle amplitude and timing differences between the ears.

Vision is even more amazing! The brain can recover three-dimensional images hidden in random-dot autostereograms ("Magic Eye" images), and it enables us to navigate through a crowded space while simultaneously recognizing faces. In such visual tasks, the brain is processing an immense flow of incoming signals from photoreceptors, about 10 megabits per second! (there is some data compression already occurring in the retina). Not surprisingly, vision consumes a significant share of brain power: estimates range from at least 20% to as much as 50% of total brain activity. The challenge of replicating human vision in artificial systems is so daunting that, in the 1980s, some researchers admitted they persisted only because humans do in fact see; otherwise, they might have concluded that vision, as we experience it, is impossible.

Inspired by nature

Reflecting on the evolution of our senses prompts us to consider the constraints that shaped them and the trade-offs that were optimized. Nature has already solved many of the same challenges we face in experimental design today: sensing across multiple physical modalities, managing noise, adaptive calibration, balancing range and resolution, optimizing sensor placement and integrating diverse data streams. Just as the brain interprets and assigns meaning to sensory input, so too must we extract knowledge from the data we collect. Let's apply the insights gained from analyzing human perception to guide our experimental strategies, using Nature's solutions to inform our own.

Exercises

The following exercises are intended to explore the subtleties of human perception. Perform each one attentively, and repeat them to refine your observations.

- Learn the steps of tasting: Look, smell, taste and reflect. Brew one cup each of English Breakfast and Earl Grey teas using the same brewing method. Take your time to carefully observe and notice the subtle differences in aroma, texture and flavor between the two teas. Tip: Cleanse your palate with plain crackers or water between tastings. Give your palate a minute or two to reset before the next sip.

- Go to a quiet natural setting. Listen. Gradually focus your attention on specific sounds: the wind, the movement of leaves and individual birds. Can you echolocate them? Notice how your auditory perception sharpens over time.
- Explore the field of view of your eyes and locate your blind spot. Where does your vision fade?
- Search for a variety of optical illusions or "Magic Eye" images. Observe how your perception can shift or be misled. What does this suggest about how your brain interprets visual information?
- Study paintings by L. da Vinci, J. Vermeer, C. Monet, S. Dalí and M.C. Escher. What do you literally see on the canvas, and what do you perceive beyond that? Consider light, form, perspective and illusion.
- Touch an orange and a rock of similar size. Pay attention to their shape (roundness, texture, smoothness), material properties (stiffness, temperature), and weight. Focus on your index finger: analyze spatial resolution (ability to distinguish fine details) and threshold pressure to detect contact.

What have you learned about the precision, limitations and adaptability of your senses? How do these sensory experiences inform your understanding of perception? How can these insights be applied when planning your next experiment?

Notes

1 I. Kant (1724–1804), German philosopher, from his *Critique of Pure Reason* (1781).
2 See essay on *Sensing the Experiment: Scientific Sensors (Section 4).*
3 Note: This section draws on well-documented literature to illustrate the range and limitations of human perception.
4 Umami is the "fifth taste": savory and meaty.
5 Resolution is defined by the various levels a sensor can distinguish. For *N*-levels, the sensor resolution in bits is such that $N=2^{\text{bits}}$.
6 See essay on *Optics: Light and Photography (Section 4).*

Essay 33

SENSING THE EXPERIMENT

Scientific sensors

Instruments extend the senses to reveal what the senses alone cannot perceive[1]

We perceive the physical world through our senses,[2] and we use sensors to extend our ability to observe states and changes during experiments. Over the past few decades, unprecedented and concurrent technological revolutions – in sensing, computer power, data transmission and storage – have transformed how we conceive experiments, how we operate them and how we process the data they produce.

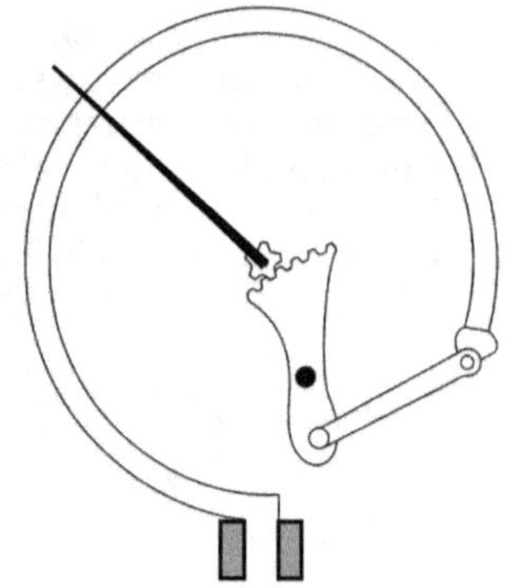

In particular, sensors have undergone dramatic changes in recent decades. Take pressure transducers: a century ago, they relied on curved Bourdon tubes that straightened as pressure increased[3]; today, they employ sub-millimeter piezoresistive or capacitive elements, often built with microelectromechanical systems (MEMS) technology. These same principles now extend to arrays that form pressure-sensitive skins and wearables with ~1 sensing element per mm^2, similar to the average mechanoreceptor density in our skin (though still two orders of magnitude lower than in our fingertips). The revolution in digital cameras, driven by consumer electronics, has been even more striking: from the first consumer models in the early 1990s to today's cell phones with 200-megapixel cameras. Modern photodetectors measure less than 1 μm, while retinal photoreceptors are 2-to-4 μm (though the combined eye–brain

DOI: 10.4324/9781003727521-37

system remains remarkably powerful). Progress is also rapid in emulating chemoreceptors, yet current electronic noses achieve only ~1 sensor per mm^2, compared with 100-to-1000 chemoreceptors per mm^2 in the nasal cavity. While these comparisons emphasize size and density, similar contrasts extend to other dimensions of performance such as range, precision, drift and durability. In every respect, advances in sensing systems have been truly remarkable.

Core components

All sensors consist of several key components, illustrated here using a classical pressure transducer as an example: (1) *a sensing element*: an elastic diaphragm; (2) *transduction into an electrical signal*: four strain gauges arranged as a full Wheatstone bridge; (3) *an input power source*: a DC voltage, typically 9 volts; and (4) *signal conditioning and output*: the voltage difference between the arms of the bridge.

Clever sensor design is evident at each stage. The sensing element can amplify the input signal through mechanisms such as geometric focusing or resonance. Transducers can be connected to cancel noise, for example, using strain gauges in a bridge configuration and differential measurements to suppress common-mode interference. Because the raw analog signal from most sensors (typically current or voltage) is vulnerable to corruption and deterioration during transmission, it is often converted into a digital output.

Some high-precision sensors, such as extensometers, pressure sensors and mass sensors, measure changes in resonant frequency: a vibrating wire (as in a guitar string) or a minute cantilever beam in MEMS (like a tuning fork). In these cases, the resonant frequency becomes the digital output.

Transduction based on energy coupling

Sensors convert various forms of energy, typically into electrical signals, including[4]:

- *Mechano-electrical*: Strains modulate resistance in strain gauges or induce a potential in piezocrystals.
- *Thermo-electrical*: Temperature changes alter resistance in thermistors or generate a potential in thermocouples.
- *Chemo-electrical*: Chemical activity is translated into electrical signals (e.g., pH meters).
- *Mechano-magnetic-electrical*: Geophones convert ground motion into voltage through the relative movement of a magnet and coil.
- *Mechano-electromagnetic*: Linear variable differential transformers (LVDT) measure displacement by modulating magnetic inductance.

- *Chemo-optical-electrical*: Changes in light absorption or emission, caused by chemical compounds, impact optical detectors and produce an electrical signal (e.g., chromatographs).
- *Mechano-optical*: Photoelastic materials experience strain-induced birefringence and, under polarized light, display fringe patterns that are associated with the applied stress field.

Know your sensors

Effective experimental design begins with a clear purpose: a precise understanding of what you intend to measure is essential for measuring it accurately – an inherent circularity in measurement. Additionally, you must understand how sensors work to select the most appropriate ones. The following are key sensor characteristics that inform sensor selection:

- *Core elements and modality*: This includes the sensing element, the transduction component, required input, type of output and the need for signal conditioning and noise control.
- *Range*: The minimum and maximum values the sensor can accurately measure.
- *Dynamic range*: The ratio between the largest and smallest detectable signals. A sensor with a high dynamic range can register both weak and strong signals without distortion or saturation.
- *Resolution*: The smallest increment in a quantity that a sensor can distinguish.
- *Precision*: The consistency of repeated measurements (i.e., the scatter around a central value).
- *Accuracy*: The proximity of the sensor output to the true value.
- *Response time*: The time required for the sensor to provide a stable value after a step input.
- *Transfer function*: The mathematical relationship between the sensor's output signal and the input. This is the primary goal of calibration.
- *Frequency response*: The amplitude and phase of a system's output relative to a steady sinusoidal input, expressed as a function of frequency. It is a specific form of the system's transfer function.
- *Hysteresis*: The output may depend on the direction of change due to factors such as stickiness, friction or slack.
- *Drift*: A gradual deviation from true values over time, often caused by internal changes like self-heating in electronics or creep in mechanical components.
- *Directivity*: The sensor's response may vary with the direction of the impinging input, as in microphones (Note: directivity can improve performance in environments with competing signals).

- *Crosstalk*: The unintended coupling of signals from adjacent sensors.
- *Sensitivity to environmental conditions – compatibility*: Sensors may be affected by pressure, temperature, humidity or chemicals. Conversely, they must comply with relevant regulations (e.g., health applications).
- *Power consumption*: This is a critical factor for sensors in hard-to-reach settings, such as implants beneath the skin or deep-seafloor monitoring nodes.
- *Longevity*: The sensor's lifespan informs installation strategies and replacement schedules; in some cases, sensors are deployed in "use-and-lose" mode.
- *Reliability and redundancy*: Critical operations may require duplicate sensors to ensure reliability and data quality.
- *Cost*: It is always a trade-off with other characteristics such as precision and longevity.

Distributed sensing and smart systems

Modern sensor technology integrates advanced capabilities such as wireless communication and on-board data processing within the sensing array itself. Another striking development is distributed sensing systems based on fiber-optic technology, where localized changes in light transmission are used to detect strain, temperature, pressure or chemical changes with meter-scale spatial resolution along a single optical fiber that can span several kilometers. These measurements exploit optical effects, including reflection, interference, diffraction, scattering and frequency shifts (e.g., Brillouin scattering). Additionally, fiber-optic sensors are inherently immune to electromagnetic interference and are well-suited for harsh chemical environments.

As sensors become more capable, smaller and more connected, they increasingly form the backbone of complex sensing systems, including sensor arrays and smart systems where measurements trigger actions in close feedback loops. Defibrillators, for instance, continuously monitor cardiac rhythms and deliver electrical shocks to restore normal function. Anti-lock braking systems (ABS) use wheel-speed sensors to detect rapid deceleration, indicating that a wheel is nearing lockup; the system then modulates brake pressure to prevent it. Modern transportation networks also rely on embedded sensors in roadways and vehicles to collect real-time data on traffic flow and environmental conditions, adjusting dynamic traffic signals to improve both efficiency and safety.

Sensors are also transforming biomechanics. Wearable devices combine accelerometers, gyroscopes and muscle-activity detectors to continuously monitor movement and posture. For athletes, these tools help optimize performance and prevent injuries. In rehabilitation, sensor-based feedback enables personalized therapy with real-time data guiding progress and adjustments.

Closing thoughts

Sensors extend the boundaries of human perception, enabling us to engage with the world in ways once unimaginable. Quietly, yet profoundly, they are reshaping industries, enhancing lives... and revolutionizing experimental research. In many ways, sensors are the cornerstone of tomorrow's breakthroughs.

Indeed, the potential of modern, sophisticated sensors is extraordinary. They deliver high-density, multi-physics data streams, enable real-time experimental control and support intelligent adaptation of boundary conditions.

Still, the responsibility lies with you – *the experimentalist* – to choose wisely and apply sensors effectively within the framework of a well-designed study. Selecting the right sensor is not merely a technical choice; it is a pivotal step in capturing the essence of the phenomenon under investigation. Ultimately, progress depends on our grasp of fundamental concepts, the clarity of our questions and our command of the technologies at hand.

Exercises

1 How would you measure the speed of light today? How was it measured in the 19th century (e.g., H. Fizeau and later A. Michelson)? And in the 17th century (e.g., Rømer's method)?
2 Bourdon tubes, known for their robustness and reliability, are still widely used to monitor local pressure in industrial applications. Design a system where a Bourdon tube serves as the sensing element, but replace the traditional gears and needle with a displacement transducer (such as an LVDT) or strain gauges to generate an electrical output.
3 How would you study the beating dynamics and synchronization of cardiomyocytes cultured in a Petri dish?
4 Modern smartphones are equipped with a wide array of sensors beyond the impressive camera: microphone, accelerometer, gyroscope, magnetometer, proximity sensor, barometer, fingerprint sensor, infrared face recognition, internal temperature sensor and humidity sensor. Investigate the characteristics of these sensors and explore related applications available online (e.g., image-processing apps, fitness trackers and navigation tools). Brainstorm opportunities where you could leverage these sensors in your own research.

Notes

1 Inspired by G. Galilei (1564–1642), Italian astronomer and physicist, reflecting on the telescope versus the naked eye.
2 See essay on *Sensing the World Around Us: Our Natural Sensors (Section 4).*
3 This sketch is based on images sourced from the internet.
4 See essay on *Electronics: Fundamentals and Applications (Section 4).*

Essay 34

ELECTRONICS

Fundamentals and applications

Atoms are not indivisible, for negatively electrified particles can be torn from them by the action of electrical forces[1]

Modern experiments increasingly depend on electronic components to observe, control and investigate physical phenomena. This brief essay presents fundamental concepts to support the understanding, design and integration of electronics in experimental research.

A useful starting analogy is to compare the flow of electrons through circuits to the flow of water molecules through pipes. Imagine a pump lifting *water* from the ocean into a storage tank, increasing its potential energy. When a valve is opened, water flows through the piping system beneath the tank, encountering various resistances on its way back to the ocean. The flow rate can be measured in liters per second, which corresponds to approximately 3.3×10^{25} water molecules per second. Similarly, in an electrical circuit, *electrons* are pumped from ground to a battery at a higher electric potential. When a switch is closed, these charges flow through a circuit, interacting with various impedances before returning to ground. Electric current $i(t)$ is measured in amperes or coulombs per second $[A]=[C]/[s]$, which is equivalent to 6.2×10^{18} electrons per second.

The design of electronic components is fundamentally based on understanding how materials respond to electric and magnetic fields. In metals, the outermost electrons are not bound to individual atoms and can move freely in response to an applied *electric field*; hence, metals are excellent *conductors*. In contrast, non-conductors or *dielectrics* typically involve covalent bonds, as in most plastics (with carbon-based compounds being a notable exception), or ionic bonds, as in dry solid salts. Interestingly, when a salt

DOI: 10.4324/9781003727521-38

dissolves in water, it dissociates into hydrated cations and anions. An externally applied voltage difference between electrodes establishes an electric field that drives these mobile ions toward oppositely charged electrodes, allowing the solution to conduct an electric current.

A material's response to an applied *magnetic field* arises from its atomic structure as well. Atomic charges in motion – through orbital movement or intrinsic spin – generate small magnetic fields. In atoms with unpaired electrons, these spins are usually randomly oriented but can align in an external field; this alignment causes a weak macroscopic response in *paramagnetic* materials and a much stronger effect in *ferromagnetic* substances due to collective spin alignment. In materials with all electrons paired, a *diamagnetic* response emerges from subtle changes in orbital motion. Finally, although the magnetic moment of protons is about three orders of magnitude smaller than that of electrons, it is still sufficient to interact with magnetic fields; this interaction forms the basis of Nuclear Magnetic Resonance (NMR) for characterizing and imaging diamagnetic materials.[2]

Resistor (dissipates energy)

The electric current $i(t)$ [in amperes A] flowing through a resistor is proportional to the applied voltage $v(t)$ [in volts V] and inversely proportional to the resistance R [in ohms Ω] according to Ohm's law:

$$i(t) = \frac{1}{R} v(t) \tag{34.1}$$

The resistance R of a resistor depends on its geometry and material properties. For a cylindrical resistor of length ℓ [m] and cross-sectional area A [m^2], made of a material with resistivity ρ [Ω m], the resistance is:

$$R = \rho \frac{\ell}{A} \tag{34.2}$$

Resistance measurements are ubiquitous in experimental research. A few key examples include:

- *Material characterization: resistivity.* Measure the resistance R of a cylindrical specimen and compute the material's resistivity ρ (using Eq. 34.2).
- *Sensors.* A conductor's electrical resistance changes when it is mechanically deformed. Strain gauges use this principle and serve as the sensing element in devices ranging from pressure transducers to load cells.

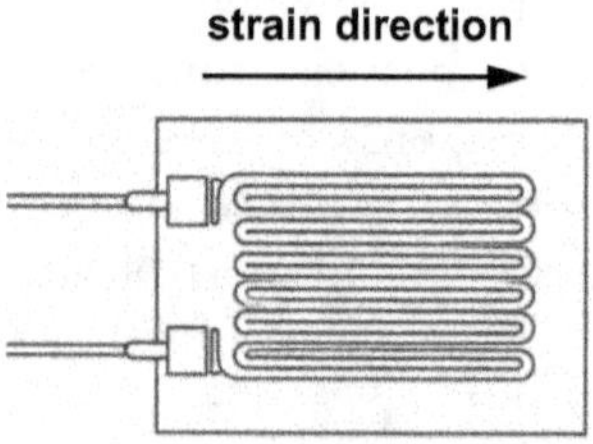

- *Electrical resistivity tomography ERT*. By applying and measuring potentials across a domain, one can reconstruct spatial variations in resistivity. ERT is used in material characterization, medical imaging and geophysical surveys.
- *Process monitoring*. Track changes such as fracture propagation, corrosion, ionic diffusion or phase changes (e.g., freezing) by monitoring resistance over time.

Capacitor (stores energy in an electric field)

A capacitor consists of two conductive plates separated by a dielectric material. When a time-varying potential $v(t)$ is applied, electrons flow and accumulate on the negative plate. Due to Coulomb repulsion, free electrons on the opposite plate are displaced, leaving it positively charged. When the potential is removed, electrons flow back. The charge $q(t)$ [in coulombs] stored in the capacitor is proportional to the applied voltage:

$$q(t) = C\,v(t) \tag{34.3}$$

where C is the capacitance [in farads $F = \text{coulomb} / \text{volt}$]. The current flowing into or out of the capacitor is the time derivative of the charge:

$$i(t) = \frac{dq(t)}{dt} = C\frac{dv(t)}{dt} \tag{34.4}$$

The amount of charge a capacitor can store increases when the gap h [m] between the plates is reduced, and when the dielectric material can polarize, thereby reducing Coulomb repulsion. The capacitance of a parallel-plate capacitor with plate area A [m^2] is given by:

$$C = \varepsilon_0 \kappa' \frac{A}{h} \tag{34.5}$$

where $\varepsilon_0 = 8.85 \times 10^{-12}\,\text{F/m}$ is the permittivity of free space and κ' [ ] is the dielectric relative permittivity (also known as the dielectric constant, though *relative permittivity* is the more accurate term). These relationships give rise to important techniques in material characterization and sensing technology:

- *Material characterization: permittivity*. By measuring the capacitance C_m of a capacitor filled with a test material and comparing it to the empty capacitor C_0 (with the same geometry), the material relative permittivity is obtained as $\kappa' = C_m / C_0$ (from Eq. 34.5).

- *Sensors: capacitive proximeters.* Since capacitance is highly sensitive to the plate gap h, capacitive sensors are widely used for high-resolution, non-contact displacement measurements.
- *Capacitive touchscreens and robotics.* These technologies detect changes in capacitance caused by the proximity of a finger or stylus, enabling precise input and feedback mechanisms.

Inductor (stores energy in a magnetic field)

An electric current i [in amperes A] flowing through a wire generates a magnetic field B [in tesla T], as described by Ampère's Law. According to the right-hand rule, if your thumb points in the direction of the current, your fingers curl in the direction of the magnetic field lines around the wire. When the wire is bent into a coil, the magnetic field lines from each loop reinforce one another, producing a strong magnetic field inside the coil. If the current varies in time $i(t)$, the magnetic field $B(t)$ also varies. According to Faraday's Law, a time-varying magnetic field induces a time-varying electric field, implying a voltage across the coil. This leads to the voltage-current relationship for an inductor:

$$v(t) = L\frac{di(t)}{dt} \tag{34.6}$$

where L is the coil inductance [in henry H], which depends on the number of turns n [ ], the coil length ℓ [m] and its cross-sectional area A [m²]:

$$L = \mu_0 \mu_r \frac{n^2 A}{\ell} \tag{34.7}$$

where $\mu_0 = 4\pi \times 10^{-7}\,\mathrm{H/m}$ is the permeability of free space, and μ_r [ ] is the relative magnetic permeability of the material inside the coil. Inductors enable a variety of applications in experimental research:

- *Material characterization: magnetic permeability.* The relative magnetic permeability $\mu_r = L_m / L_0$ of a material can be found by measuring the inductance of a coil filled with the material L_m and comparing it to the inductance of the coil in air L_0 (from Eq. 34.7).
- *Transformers.* A time-varying voltage applied to a primary coil creates a changing magnetic field that induces voltage in a secondary coil wound around a common magnetic core. Transformers are essential in voltage regulation, circuit isolation and wireless chargers.

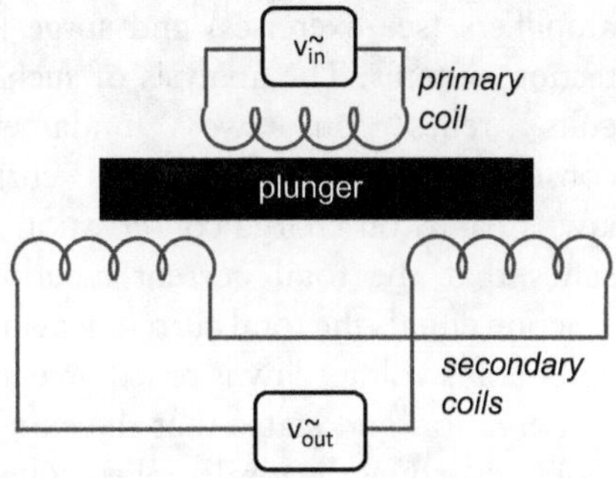

- *Sensors*. Inductive proximeter sensors generate oscillating magnetic fields and detect nearby metal objects by changes in inductance. The linear variable differential transformer (LVDT) is a widely used displacement transducer based on transformer principles.
- *Actuators and motors*. Solenoids convert electrical energy into mechanical force, enabling linear motion in devices like valves and loudspeakers, or rotational motion in electric motors. The reverse process, where mechanical motion generates electrical signals, is used in devices such as geophones.
- *Charge trajectory control*. Moving charges are affected by magnetic fields in addition to electric fields (the Lorentz force). This is the basis for mass spectrometry and scanning electron microscopy (SEM).[3]

Impedance – Circuit analysis

Let's use the general term impedance z [in ohms Ω] to describe the ratio between the voltage $v(t)$ and the current $i(t)$ across each type of circuit element. Impedance generalizes resistance R to include the frequency-dependent behavior of capacitors and inductors. From Eqs. 34.1, 34.4 and 34.6, we can define the impedance for each circuit element assuming sinusoidal signals of angular frequency ω[4]:

Impedance	*Resistor*	*Capacitor*	*Inductor*
$z = \frac{v(t)}{i(t)}$	$z_R = R$	$z_C = \frac{-j}{\omega C}$	$z_L = j\omega L$

Note that the impedance of a capacitor $z_C \to 0$ as $\omega \to \infty$, meaning that capacitors allow high-frequency signals to pass through while blocking low-frequency signals. On the other hand, the impedance of an inductor $z_L \to 0$ *as* $\omega \to 0$, so inductors allow low-frequency signals to pass while impeding high-frequency components.

These expressions allow us to analyze circuits using the efficiency of "complex algebra", which is not all that complex! When combined in circuits, the frequency-dependent behavior of resistors, capacitors and inductors enables the design of analog filters (low-pass, high-pass and band-pass), resonant circuits, amplifiers (see exercises) and surge protection systems. The analysis of such circuits relies on two fundamental conservation laws. Kirchhoff's current law is based on charge conservation and states that the total current entering a junction equals the total current leaving it. Kirchhoff's voltage law is based on energy conservation and states that the sum of voltage drops Δv_i around any circuit loop must equal the total applied voltage v_{in}, provided by a signal generator SG.

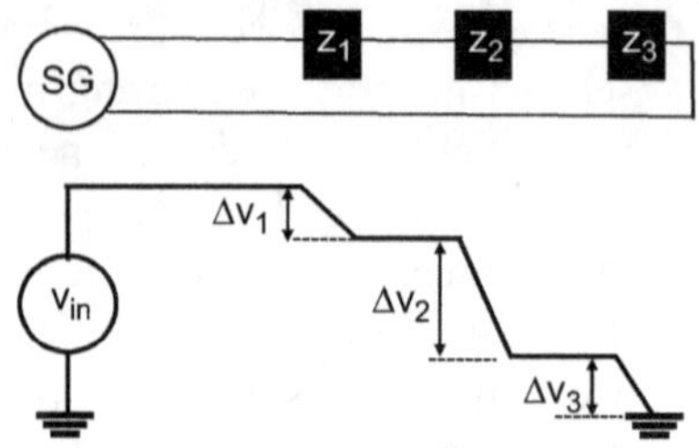

These laws allow us to calculate the equivalent impedance z_{eq} of two components with impedances z_1 and z_2, depending on whether they are connected in series or parallel:

Parallel: (currents add - 1st law)

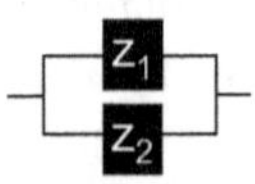

$$\frac{1}{z_{eq}} = \frac{1}{z_2} + \frac{1}{z_1}$$

Series: (voltage drops add - 2nd law)

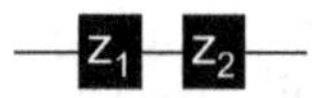

$$z_{eq} = z_1 + z_2$$

Basic electronic devices

Depending on the application, you will need various peripheral electronic devices. Commonly used instruments include:

- *Multimeter*: Measures voltage, current and resistance. It features high input impedance when measuring voltage (to avoid loading the circuit) and very low input impedance when measuring current (to ensure accurate current flow measurement).
- *Oscilloscope*: Visualizes time-varying electrical signals and is essential for analyzing waveforms and debugging circuits. It can display signals with their inherent DC offset (DC coupling) or with the DC component removed (AC coupling).
- *Function Generator*: Produces standard electrical waveforms such as sine, square, triangle, pulse, noise, arbitrary waveforms, and frequency sweeps. These outputs are typically low power, so a power amplifier may be required to drive loads like actuators or shakers.
- *Pre-Amplifier and Filter*: Often placed between sensors and the oscilloscope, this device amplifies low-level signals and filters out noise before digitization.
- *Data Acquisition System DAQ*: Captures analog or digital signals for storage, analysis and post-processing.
- *Power Supply (DC or AC)*: Provides controlled voltage or current to power circuits and sensors. Adjustable supplies allow precise input control during testing or development.

In most setups, these devices share a common ground to minimize electrical noise. Always ensure that measurement instruments – especially oscilloscopes – are connected correctly to avoid introducing ground loops or short circuits. A helpful practice is to "unfold" the circuit diagram to visualize current paths and voltage drops clearly. If grounding is a concern, consider using floating-ground measurement instruments to isolate the measurement system from the circuit under test.

Go ahead... thoughtfully

Mastering the concepts presented in this essay will not make you an electrical engineer, but it will give you a solid foundation in electrical measurements and the vocabulary needed to communicate effectively with experts in the field. So go ahead, explore the world of electrical measurements. Start basic, take your time, keep learning and stay curious. Remember: electricity does not forgive mistakes. Always double-check your connections. Respect voltage. Safety and patience are just as important as creativity.

Exercises

1 Study the inner workings of common electronic systems: a switch, a circuit breaker, a ringer circuit from a vintage phone, and a DC motor. *Follow the charges* and apply concepts from this essay to analyze and understand these systems.

2 Visit research laboratories and observe the instrumentation used in different experiments. Identify the electronic devices, their operating frequencies, noise-control strategies, cables and connectors, and impedance-matching techniques. Consider how these elements affect signal integrity and measurement quality.

3 Add the impedances of a resistor R, a capacitor C and an inductor L in series. Plot the real and imaginary parts of the equivalent admittance $Y = 1/z$ *vs.* frequency. Use: $L = 10\text{mH}, C = 2.53\mu\text{F}, R = 3.1\Omega$. Compute the resonant frequency: $f_o = \left(2\pi\sqrt{LC}\right)^{-1}$ and the quality factor $Q = \frac{1}{R}\sqrt{L/C}$.

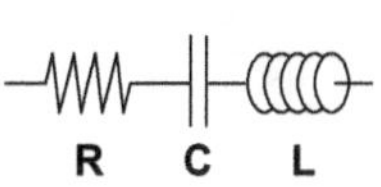

4 Reproduce the RLC circuit on a breadboard. Use a function generator and a two-channel oscilloscope to study the response. Channel 1: measure the input voltage $v_0(t)$ across the full circuit. Channel 2: measure the voltage drop $\Delta v(t)$ across the resistor to compute the current $i(t) = \Delta v(t)/R$. Caution: Make sure the oscilloscope ground clips are connected to the common circuit ground to avoid ground loops or shorts.

5 Strain gauges. Analyze the 3-wire configuration of an external quarter-bridge. Demonstrate that it eliminates the effects of lead resistance variations when the bridge is balanced.

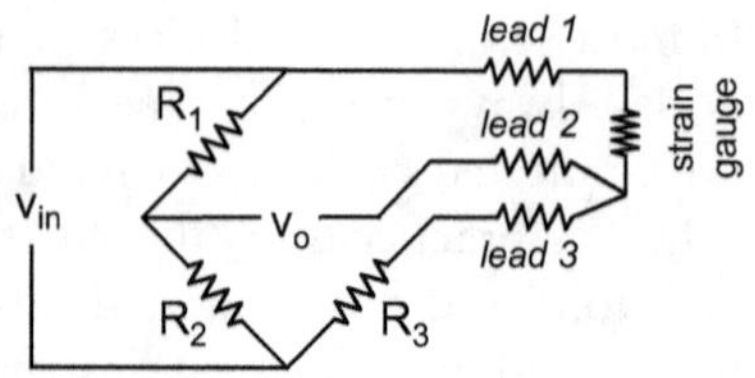

6 Instrument a square rod with four strain gauges (one per face). Specify installation and wiring to enable the rod to function as a longitudinal load cell, a flexural load cell or a torsional load cell. *Think carefully*!

Notes

1 J.J. Thomson (1856–1940), British physicist who discovered the electron.
2 See essay on *Advanced Characterization Systems (Section 4)*.
3 See essay on *Advanced Characterization Systems (Section 4)*.
4 Hint: replace the time varying current or voltage with a complex exponential $e^{j\omega t}$ where j is the imaginary unit.

Essay 35

WAVES

Carriers of energy and information

If you want to find the secrets of the universe, think in terms of energy, frequency and vibration[1]

Waves are messengers that carry *information and energy* across space and time:

- *Mechanical waves* allow us to hear sounds and even break kidney stones. They travel through a medium, causing a material element to vibrate either in the same direction as the wave's propagation (known as P-waves) or transversely to it (shear S-waves).
- *Electromagnetic waves* enable us to see and to heat food in a microwave oven. They consist of time-varying electric and magnetic fields that oscillate transversely to both the direction of travel and to each other.

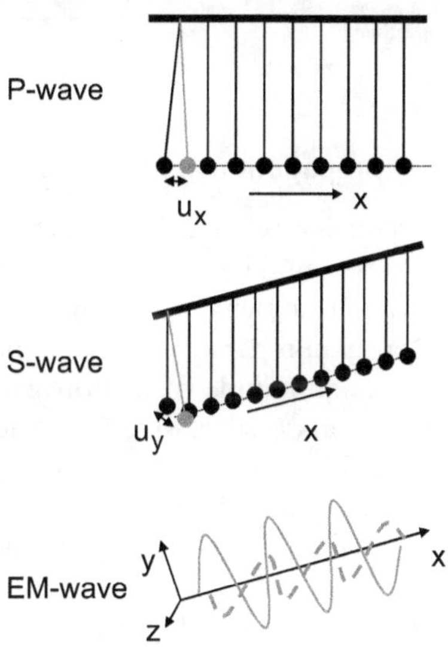

Most experiments use waves as information carriers. A wave train enters a medium with a given amplitude-phase spectrum and emerges with a transformed spectrum. This transformation encodes valuable information about the medium, which is exploited in wave-based material characterization, advanced imaging techniques such as ultrasound diagnostics, and probing the Earth's internal structure. In other applications, waves are deliberately

DOI: 10.4324/9781003727521-39

modulated to encode information; modulation may alter amplitude (AM radio), frequency (FM radio) or phase (digital holography and imaging systems), or it can be structured into pulses (digital communications and fiber optics). In every case, the wave acts as the carrier, while the modulation encodes the information.

The spectrum: names and applications

The chart below summarizes the spectra of mechanical and electromagnetic (EM) waves, along with common applications. Note that the term *sound* has a dual meaning that can be a source of confusion: in everyday use, it refers to what we hear (the 20 Hz to 20 kHz gray band shown below), whereas in physics, it more broadly denotes propagating mechanical vibrations (both P- and S-waves).

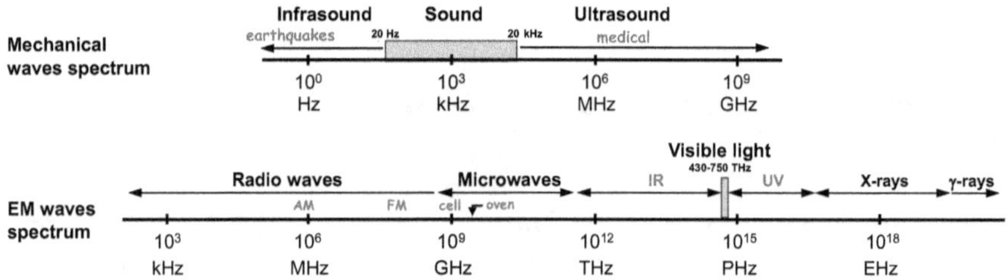

Key propagation characteristics

Although mechanical and electromagnetic waves arise from different physical processes and interact with media in distinct ways, they share many common wave phenomena, including reflection, refraction and diffraction.[2] Key wave propagation parameters are summarized in the table below.

The amplitude of a propagating one-dimensional plane wave decreases due to material losses, quantified by the spatial attenuation coefficient α $[m^{-1}]$. The wave amplitude decays exponentially with distance x, reaching $A_x = A_0 / e$ at a distance $x = 1/\alpha = \beta$ known as the skin depth. In addition to material attenuation, partial reflections at interfaces, where the transmitted amplitude is reduced by a factor of $(1 - R)$, and geometric spreading in cylindrical or spherical waves further diminish the amplitude as the wave propagates.

Media with internal time scales exhibit frequency-dependent propagation. For example, during electromagnetic wave propagation in water, molecules can rotate to align with the oscillating electric field up to ~10 GHz; at higher frequencies, molecular orientation cannot keep pace with the rapidly oscillating field (Debye relaxation). Similarly, in saturated porous media, water moves with the solid skeleton during low-frequency P-wave propagation, but

	Mechanical waves	*Electromagnetic waves*
Starting point	$F = \text{ma}$	Maxwell's Equations
Wave equation	$M\frac{\partial^2 u_\parallel}{\partial x^2} = \rho\frac{\partial^2 u_\parallel}{\partial t^2} + \eta_p\frac{\partial}{\partial t}\left(\frac{\partial^2 u_\parallel}{\partial x^2}\right)$ $G\frac{\partial^2 u_\perp}{\partial x^2} = \rho\frac{\partial^2 u_\perp}{\partial t^2} + \eta_s\frac{\partial}{\partial t}\left(\frac{\partial^2 u_\perp}{\partial x^2}\right)$	$\frac{\partial^2 E}{\partial x^2} = \mu\varepsilon\frac{\partial^2 E}{\partial t^2} + \mu\sigma\frac{\partial E}{\partial t}$ $\frac{\partial^2 B}{\partial x^2} = \mu\varepsilon\frac{\partial^2 B}{\partial t^2} + \mu\sigma\frac{\partial B}{\partial t}$
Material parameters	M and G: constrained and shear moduli ρ: mass density η_p and η_s: longitudinal and shear viscosity	μ: magnetic permeability $\varepsilon = \varepsilon_0\kappa'$ dielectric permittivity κ': relative permittivity σ: electrical conductivity
Velocity	$v_P = \sqrt{\frac{M}{\rho}} = \sqrt{\frac{B + \frac{4}{3}G}{\rho}}$ and $v_s = \sqrt{\frac{G}{\rho}}$ P-wave: requires bulk stiffness $B \rightarrow M$ S-wave: requires shear stiffness G	$v = c_0\left\{\frac{1}{2}\left[\sqrt{\kappa^2 + \left(\frac{\sigma}{\varepsilon_o\omega}\right)^2} + \kappa\right]\right\}^{\frac{-1}{2}}$ for $\mu = 0$ ε_0: vacuum permittivity; ω: angular frequency $c_0 = 3\times10^8\ \text{m/s}$: speed of light in free space
Implications	P and S-waves cannot propagate in free space S-waves cannot propagate in fluids	They can propagate in free space $\mu = 0\ \kappa = 0$ (hence, we can see the universe)
Amplitude decay with distance x	$A_x = A_0\,e^{-\alpha x}$ for particle motion $u_\parallel$ or $u_\perp$ $\alpha = \frac{\omega^2\eta}{2\rho v^3}$ (both P and S)	$A_x = A_0\,e^{-\alpha x}$ for field amplitude E or B $\alpha = \frac{\omega}{c_o}\left\{\frac{1}{2}\left[\sqrt{\kappa'^2 + \left(\frac{\sigma}{\varepsilon_o\omega}\right)^2} - \kappa\right]\right\}^{\frac{1}{2}}$ for $\mu = 0$

(Continued)

	Mechanical waves	*Electromagnetic waves*
Impedance	P-waves: $z_P = \rho v_P$ S-waves: $z_s = \rho v_s$	$z = \sqrt{\mu / \varepsilon}$
Reflection	$R = \frac{z_2 - z_1}{z_2 + z_1}$ $R = A_r / A_i$: ratio of reflected amplitude A_r to the incident amplitude A_i Both equations assume normal incidence and lossless media	
Refraction (Snell's law)	$\frac{\sin\theta_i}{v_1} = \frac{\sin\theta_t}{v_2}$ θ_i: angle of incidence; θ_t: angle of transmission (refraction)	
Diffraction	Diffraction occurs at edges It is strong when the wavelength is comparable to the obstacle size Analysis: Every point on a wavefront acts as a secondary source of Huygens wavelets	

the two decouple at high frequencies (Biot relaxation). In both cases, the propagation velocity depends on frequency, leading to *dispersion*. As a result, the angle of refraction also becomes frequency-dependent according to Snell's law, causing a signal to separate into its constituent frequencies. The same physics explains the familiar dispersion of white light into a spectrum of colors when it passes through a Newton prism.

While mechanical and electromagnetic waves share many analytical similarities, they also exhibit important phenomenological differences. For instance, a P-wave striking an inclined boundary generates not only reflected and refracted P-waves but also reflected and refracted S-waves – a process known as *mode conversion*, which arises from the equilibrium and compatibility of deformation at the interface. In contrast, an electromagnetic wave incident on a boundary does not convert into other wave types but may undergo polarization-dependent reflection and transmission (e.g., at the *Brewster angle*).

Specimen size, wavelength and resolution

An elastic wave traveling along a material rod or an electromagnetic wave traveling along a waveguide is at resonance when the wave reflecting from both ends reproduces itself after a round trip. The first resonant mode defines the longest wavelength, $\lambda_{max} = v / f$, that can be studied with a given specimen size L. Because reflection depends on the impedance at the boundaries, the resonance condition does as well: the longest wavelength is $\lambda_{max} = 2L$ for free-free ($z = 0$ at both ends) and fixed-fixed boundaries ($z = \infty$ at both ends), whereas $\lambda_{max} = 4L$ for a free-fixed condition ($z = 0$ at one end and $z = \infty$ at the other).

Consequently, high-frequency tests ($f > v / \lambda_{max}$) are analyzed using wave propagation theory, whereas low-frequency tests ($f < v / \lambda_{max}$ and $\lambda \gg L$) are modeled using lumped-parameter systems, such as mass-spring-dashpot systems for mechanical perturbations or capacitors-inductors-resistors[3] for electromagnetic studies. For reference, a $\lambda = 300\,\text{mm}$ wavelength in air corresponds to 1.14 kHz acoustic waves $(v_P = 343\,\text{m/s})$ and 1 GHz electromagnetic waves $(v = c_0 = 3\times10^8\,\text{m/s})$.

Spatial resolution scales with wavelength and therefore improves with increasing frequency; however, skin depth decreases at higher frequencies (see the table). The operating frequency is therefore selected to balance resolution and penetration. For example, ultrasonic imaging in medical diagnostics operates at 2 MHz $(\lambda \approx 0.7\,\text{mm})$ to achieve deeper penetration and at 18 MHz $(\lambda \approx 0.09\,\text{mm})$ to improve resolution, albeit with reduced penetration depth.

Take-home messages

Wave propagation phenomena are remarkably rich, playing an outsized role in experimental research... and in our daily lives! Experiments must start with well-defined boundary conditions, and data analysis must consider the full range of wave phenomena. Finally, remember that every wave has its own length scale λ and time scale $T = 1/f$, making it an ideal tool for probing the spatial and temporal structure of a medium.

Exercises

1 Consider a 100 mm-long water specimen. What is the lowest excitation frequency that can be analyzed as wave propagation for both mechanical and electromagnetic waves?
2 Analyze the size of the human eye and ear in relation to the wavelengths of visible and audible waves, respectively. What can you conclude? Why might our sensors have evolved to these particular sizes?
3 Explore the use of *ring-down* methods for material characterization (based on the amplitude decay of a signal in a resonant system). Compare their use in mechanical systems (e.g., resonant column tests) and electromagnetic systems (e.g., optical cavity ring-down spectroscopy).
4 Connect a small microphone to an oscilloscope (a preamplifier may be required), and connect a small speaker to a signal generator through a T-connector so that part of the generator output is sent to the second oscilloscope channel as a reference signal. Surround the setup with sound-absorbing materials to minimize reflections and background noise. (a) Vary the input frequency and record the detected amplitude to plot the frequency response of the speaker-microphone pair. (b) Record and analyze the characteristic frequency range of your voice. (c) Determine the *frequency range* you can hear. (d) Measure the *speed of sound* in air using time-delay or wavelength-based methods. (e) Fix the source frequency at 4 kHz and gradually move the microphone away from the speaker to identify the transition from *near-field* to *far-field* behavior. (f) Keep the microphone at a fixed radial distance from the speaker and rotate it around in small angular steps; plot amplitude vs. angle to determine the speaker's *directivity* pattern.
5 Prepare two identical 25 cm-long antennas (quarter-wavelength monopoles for 300 MHz). Using coaxial cables, connect one antenna to an oscilloscope and the other to a low-power signal generator; use a T-connector to feed part of the generator output to the second oscilloscope channel as a reference. (a) Verify impedance matching between components. (b) Measure and characterize background noise. (c) Record amplitude as a function of distance to identify *near-field* and *far-field* regions. (d) Rotate the receiving antenna relative to the transmitting antenna in small steps to measure how *polarization* alignment affects the received amplitude.

Notes

1 N. Tesla (1856–1943), Serbian-American inventor.
2 See essay on *Optics: Light and Photography (Section 4)*.
3 See essay *Electronics: Fundamentals and Applications (Section 4)*.

Essay 36

OPTICS

Light and photography

Light makes photography. Embrace light.
Know light ... and you will know the key to photography[1]

Visible light occupies a narrow portion of the electromagnetic spectrum, yet it allows us to perceive the world in vivid detail and in colors ranging from violet $(\lambda = 0.38\mu m)$ to red $(\lambda = 0.70\mu m)$.

Optics, the science of light, has profoundly influenced the history of science and technology. In 1609, G. Galilei developed the telescope and provided observational proof of Copernicus' heliocentric model. In the 1670s, A. van Leeuwenhoek used a single-lens microscope to discover microorganisms, giving rise to the field of microbiology. The Michelson-Morley interferometry experiment of 1887 failed to confirm the existence of the luminiferous ether, paving the way to Einstein's theory of special relativity in 1905. Today, modern optical devices play a central role in experimental research, from small-scale laboratory tests and advanced analytical tools to the exploration of the universe.

The operation of optical systems is based on the wave nature of light and the associated phenomena. Propagation *velocity and attenuation* are sensitive to the properties of the medium (laser-based air pollution measurements). *Reflection* depends on impedance mismatch at interfaces (satellite imaging). *Refraction* results from differences in propagation velocity between adjacent media (lenses in microscopes and telescopes). *Diffraction* is the spreading of light when it encounters obstacles or narrow openings comparable in size to the wavelength (diffraction gratings in spectrometers). *Scattering* redirects light when it strikes small particles in the medium (particle sizing). Constructive and destructive *interference* produces alternating light and dark

DOI: 10.4324/9781003727521-40

bands through wave superposition (Michelson interferometer). *Polarized light* vibrates in only one direction (microscopy and glare control). And *dispersion* occurs when light splits into its component colors as it passes through a medium because the refractive index depends on wavelength (as in a spectrometer).

Photography

Light allows us to see, but we do not always observe...[2] Photography, like sketching, helps focus our attention and intention, stimulating interpretation and new understanding. Furthermore, photography can reveal what may escape our perception by extending our ability across wider time and spatial scales.

The story of photography begins in the 5th century BCE with the camera obscura, a dark chamber that projects images through a small aperture. Refined by I. al-Haitham (965–1039) and enhanced with lenses in the 14th century, it remained primarily a tool for studying optics until the 1820s–1830s, when photosensitive silver compounds enabled the permanent recording of black-and-white images. Color photography arrived with the Autochrome process in 1907. The digital revolution began with the development of the charge-coupled device (CCD) sensor at Bell Labs in 1969, which converted incoming photons into electrons; by the 1990s, digital cameras had already begun to replace film.

Photography has supported scientific inquiry since its early days, with astronomical photography dating back to the 1840s. Today, digital imaging has made photography widely accessible by combining advanced capabilities into compact, affordable devices.[3] Let's dive into foundational concepts and techniques that can open new doors for experimental research.

It's all about light

Photography is critically dependent on illumination and exposure control. In general, soft, diffuse yet directional *illumination* is considered "good light" because it creates gentle shadows that enhance depth perception. While diffuse light flatters subjects, hard light brings out texture and detail. To control shadows, use a main light source paired with a fill light. Adjust the white balance setting to maintain accurate colors. For specialized applications, use UV light and fluorescent dyes to observe flow or polarized light to analyze crystals or tissue. Be mindful of unwanted distortions and reflections (whether the lens or even yourself!). Finally, ensure stable illumination for long-duration photographic studies, such as time-lapse photography.

Three interrelated settings control the amount of light that reaches the film or photosensor in a camera, together determining proper *exposure*.

(1) *Aperture* is the lens opening that lets light pass through, expressed in f-stops; a wide aperture with a lower f-number (e.g., f/1.8) produces a brighter image but with a blurry background, while a narrow aperture with a higher f-number (e.g., f/16) results in a darker image but greater depth of field, keeping everything in focus. (2) The *shutter speed* controls how long the film or sensor is exposed to light; a slow shutter speed (e.g., 1/30 sec) allows more light but can cause motion blur, whereas a fast shutter speed (e.g., 1/1000 sec) freezes motion but lets in less light. (3) *ISO sensitivity* refers to the sensor or film's sensitivity to light; ISO 100 is less sensitive but produces a cleaner image, while ISO 3200 is highly sensitive but often results in grainier textures.

Photographic techniques

Photography offers a versatile range of techniques that can support and enhance experimental research:

- *Long exposure* keeps the shutter open for an extended period, allowing moving objects such as cars or stars to leave light trails, thus capturing motion and time within a single photograph.
- *Double exposure* overlays two images on the same frame to illustrate change over time, either by exposing the same section of film twice or layering two images digitally.
- *Movies or motion pictures* rely on the phenomenon of persistence of vision, in which the eye retains an image for about 1/16th of a second, allowing a rapid sequence of still images to appear as continuous motion. Animation typically uses 12 frames per second (fps), silent films ~16 fps and modern films run at 24 fps (with up to 60 fps for sports). A common artifact to watch for is aliasing, where motion appears distorted due to insufficient temporal sampling, such as wheels appearing to spin backward in a moving car.[4]
- *Stroboscopic photography* uses high-frequency flashes to capture still images of rapidly moving objects. Each burst freezes a moment in time, creating a trace of the object's trajectory. Advanced systems use signal stacking to combine multiple motion phases into a single composite image.
- *High-speed photography* obtains thousands or even millions of frames per second, allowing us to observe fast events like water droplets splashing, hummingbirds hovering, or explosive chemical reactions. Innovations like femto-photography even enable the visualization of light itself, capturing photons as they travel through objects.
- *Time-lapse photography* records frames at set time intervals and plays them back at higher speeds to visualize slow processes such as the motion of the Milky Way, the construction of buildings or the growth of cultures in a petri dish.

- *Microphotography* captures highly magnified images using optical microscopes or scanning electron microscopy (SEM), revealing details invisible to the naked eye.[5]
- *Hyperspectral spectroscopic imaging* extends human vision by recording light across a broad range of wavelengths, often beyond the visible spectrum. Each image encodes both spatial and spectral data, enabling the analysis of chemical composition, moisture content or surface changes.
- *Stereoscopy* mimics human binocular vision by acquiring two slightly offset images of the same scene. When each image is presented separately to one eye, the brain fuses them into a single three-dimensional perception, as in anaglyphs or 3D movies viewed through filtering glasses. Stereophotogrammetry further enables the extraction of 3D coordinates from stereo images taken from different angles.
- *Photoelasticity* uses transparent, birefringent materials that change their optical properties under stress. When placed between polarizing filters, these materials display colorful interference fringes that reflect internal stress fields. These fringe patterns are analyzed to quantify the stress field. Photoelastic coatings apply the same principle to real-world components by bonding a birefringent layer directly to a structure, allowing stress patterns to be observed and recorded on the actual object under load.
- *Interferometry* splits a beam of light, reflects it along separate paths and then recombines it. The resulting interference patterns reveal variations in surface height, thickness or structure with extremely high precision.
- *Shadowgraphy* visualizes changes in transparent media – such as air – by projecting light through the medium: regions with varying refractive indices create "shadows" on a screen. This technique can be used to study heat flow or standing waves. A more advanced implementation known as the Schlieren method enhances sensitivity and detail.

Post-processing

Digital image analysis further expands the reach of photography. Classical *digital image processing* techniques can be used to reduce noise, adjust contrast, sharpen or smooth edges and extract key features. More recent *AI-based algorithms*, trained on millions of related images, can reconstruct missing or degraded details with *super-resolution* capabilities. By comparing successive images, *digital image correlation* can generate a displacement map. *Motion amplification*, such as Eulerian video magnification, reveals subtle movements by detecting and amplifying pixel-intensity changes over time, making it possible to visualize building vibrations or a person's pulse. *Focus stacking* combines multiple images taken at different focal distances to produce a single image with an extended depth of field. Finally, *high dynamic range* imaging merges photographs taken at different exposures to capture a wider range of brightness levels, preserving both shadow and highlight detail in high-contrast scenes.

Lights... camera... action!

Photography can be much more than just a record of an experimental setup: it can be the very core of the experiment itself. Photography sharpens our attention, extends our perception and reveals phenomena that lie beyond plain sight. Approach photography with fresh curiosity: look more closely, more slowly and from perspectives you have not considered before.[6] Explore new techniques ... observe with intention ... and let the lens lead you to new insights.

Exercises

1 Explore perception biases through optical illusions, color tests, perspective (e.g., M.C. Escher's artworks) and hidden images in famous paintings.
2 Investigate why car headlights appear dimmer or less effective on wet roads. Why do the streetlight reflections in V. Van Gogh's 1888 *Starry Night Over the Rhône* appear as elongated lines?
3 Analyze experimental results obtained with photographic techniques, such as videos created using high-speed and time-lapse photography. Appreciate processes that are otherwise invisible to the eye.
4 Consider your field of interest. Design three experiments that use optical measurement techniques. For each experiment, include a sketch of the setup, identify the key optical components, and explain the measurement objective. Emphasize practicality, robustness, and creativity.
5 Use an optical microscope to observe various materials (e.g., salt grains, hair). Quantify surface features such as shape and roughness. How might these microscale characteristics influence the material's macroscale behavior?
6 Use a cellphone camera in close-up mode to monitor bacterial cultures growing on agar plates. Carefully design the lighting setup to minimize distortions, reflections and glare and select appropriate exposure settings. Capture detailed images in time-lapse mode, then post-process them to analyze colony morphology, size, color and growth patterns over time.

Notes

1 G. Eastman (1854–1932), American innovator, revolutionized photography by introducing flexible roll film and the Kodak camera.
2 Paraphrasing Sherlock Holmes: "You see, but you do not observe" – A.C. Doyle (1859–1930).

3 See essay *Sensing the World Around Us: Our Natural Sensors* (Section 4).
4 See essay on *Introduction to Data Processing and Modeling* (Section 5).
5 See essay on *Advanced Characterization Systems* (Section 4).
6 Paraphrased from A. Morell (1948–), Cuban-American photographer best known for his innovative use of the camera obscura technique: "I started making photographs as if I were a child myself. This strategy got me to look at things around me more closely, more slowly, and from vantage points I hadn't considered before".

Essay 37

ADVANCED CHARACTERIZATION SYSTEMS

The experimenter who does not know what he is looking for will not understand what he finds[1]

Much transformative research can be achieved with basic tools and careful observation.[2] Today's scientists and engineers also have access to instruments of remarkable sophistication, enabled by revolutionary advances over recent decades, as discussed in earlier essays.[3] This essay unveils the physical elegance underlying the core components of modern instrumentation and then examines analytical devices commonly used in research laboratories[4]

Core components

Laser = Light Amplification by Stimulated Emission of Radiation

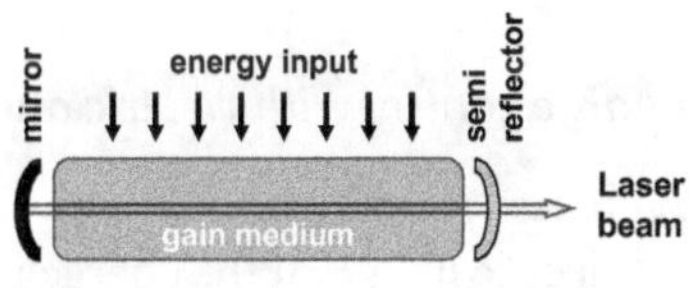

A laser operates on the principle of stimulated emission: atoms or molecules in a gain medium are first excited by an external energy source. These excited particles emit photons that share the same wavelength and phase as incoming photons. The light bounces back and forth between two mirrors in an optical cavity, becoming increasingly amplified with each pass, until one partially reflective mirror allows a coherent (in-phase), monochromatic (single-wavelength) and collimated (parallel-ray) beam to emerge.

DOI: 10.4324/9781003727521-41

Fiber optics

Light can be transmitted over long distances with minimal loss through thin, flexible glass or plastic fibers. Each fiber consists of a high-refractive-index core surrounded by a lower-index cladding, which ensures total internal reflection and keeps the light confined even when the fiber bends. Light is typically injected into the fiber using laser diodes or LEDs. A key advantage of fiber optics for data transmission is its immunity to electromagnetic interference and crosstalk, as well as its robustness in harsh environments.

Michelson interferometer

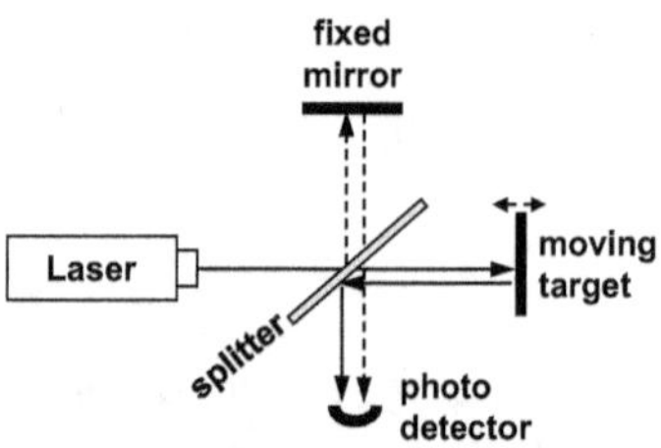

Newton's rings in thin oil films on water are a classical demonstration of constructive and destructive interference in optics. In a Michelson interferometer, a coherent light source (typically a laser) is split into two beams that travel along separate paths: one reflects from a mirror at the target and the other from a fixed reference mirror. The two reflections recombine on a screen or photodetector, producing interference fringes that shift with minute changes in optical path length. Because the target displacement δ affects the travel path twice, the phase change $\Delta\theta$ is given by $\Delta\theta / 2\pi = 2\delta / \lambda$ so the displacement can be obtained as $\delta = \dfrac{\Delta\theta}{4\pi}\lambda$. Modern electronic phase detection can resolve changes smaller than 10^{-3} radians, allowing visible-light interferometers $(\lambda \approx 500\,\text{nm})$ to achieve displacement resolution on the order of 0.4 Å, which is smaller than the diameter of an atom.

Charge control with electric and magnetic fields

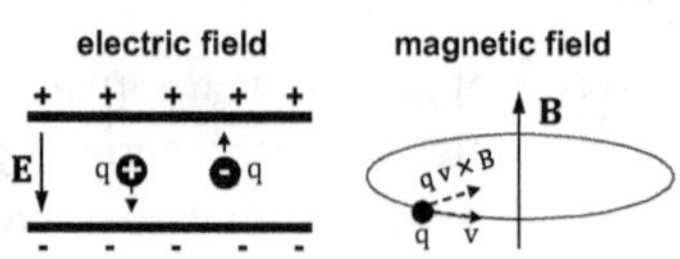

A charge q experiences a force when placed in an electric field E. If the charge is moving with velocity v, it also experiences a force in a magnetic field B. The combined effect is described by the Lorentz force $F = q\,E + q\,v \times B$.

Electron gun

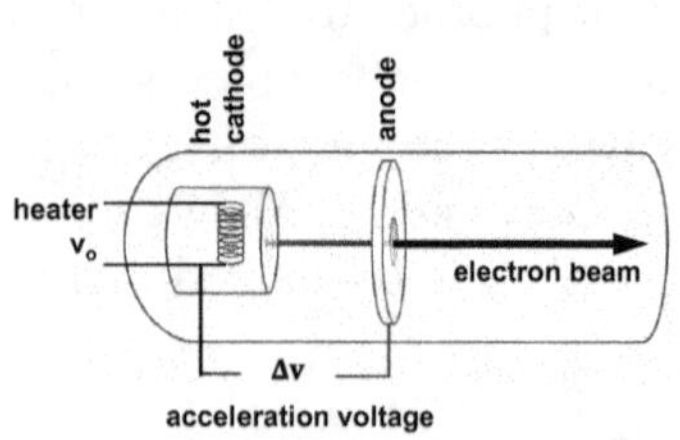

Electrons emitted from a heated metal filament, typically tungsten, are accelerated toward an anode by applying a potential difference Δv of tens to hundreds of kilovolts. The cathode-anode gap is only a few millimeters wide, producing an electric field exceeding $E = \Delta v / \text{gap} > 10^6\,\text{V}/\text{m}$ (Note: electron guns must operate under high vacuum). The resulting electron beam passes through a central aperture and is focused by controlling the Lorentz force through carefully adjusted electric and magnetic fields.

X-ray source

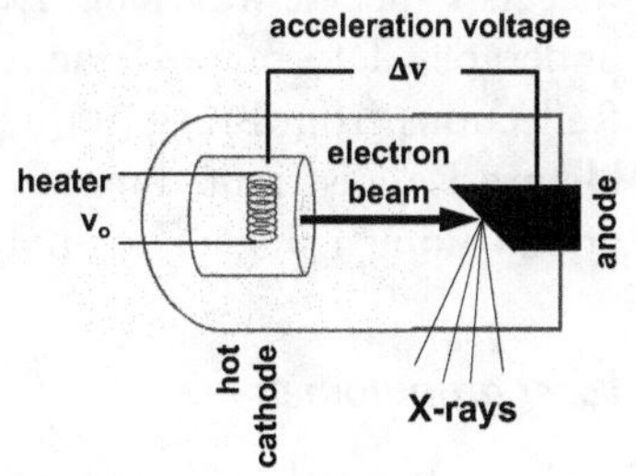

When high-energy electrons, accelerated by a potential difference $\Delta v = 30 - \text{to} - 300\text{kV}$, strike a metal target such as tungsten, molybdenum or copper located a few millimeters away, its kinetic energy is converted into X-ray radiation during rapid deceleration (bremsstrahlung radiation). Only a small fraction of the electron energy is converted into X-rays, while most is dissipated as heat; therefore, efficient cooling systems are essential.

Piezoelectric position control and vibration generation

Piezoelectric materials convert an applied voltage into mechanical deformation, typically on the order of ångströms per volt. This property enables precise nanoscale motion control for quasi-static positioning using piezoceramics, as in atomic-scale microscopy. The piezoelectric effect is also employed in dynamic applications, such as high-precision quartz crystal microbalances, where changes in resonant frequency allow detection of minute variations in mass.

Advanced data processing

A central component of all advanced instrumentation is extensive data processing. Common approaches include (1) frequency decomposition and spectral analysis, as in infrared absorption measurements[5]; (2) inversion of boundary data to reconstruct internal spatiotemporal properties, as in tomographic imaging[6]; (3) dynamic control of excitations and boundary conditions, as in scanning microscopy[7]; and (4) the use of artificial intelligence to interpret results within large database frameworks, such as in DNA sequencing.

Advanced characterization methods

The following characterization techniques illustrate how the core components described above are integrated into systems capable of probing structure and composition across scales, from atomic to geological.

Distributed fiber-optic sensing (DFOS)

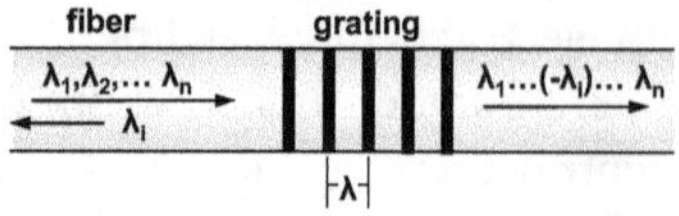

Beyond data transmission, fiber optics can also function as sensors. For example, a few-millimeter-long periodic modulation of the fiber's refractive index (a Bragg grating) reflects a specific wavelength λ; the reflected wavelength shifts when the fiber undergoes deformation due to mechanical stress or temperature variations. Reflections from Bragg gratings, along with other scattering processes, enable precise, distributed monitoring of strain and temperature at hundreds of points along a single fiber, potentially over distances of several kilometers.

Laser granulometry

Light scatters when it encounters a particle; the scattering pattern depends on particle size relative to the wavelength and on composition. Laser granulometry uses scattering to measure the size distribution of powders, suspensions or aerosols, typically for particles comparable in size to the wavelength (Mie scattering theory).

Interferometry

Interferometry is a key technique in remote sensing. For instance, interferometric synthetic aperture radar (InSAR) can assess topography and ground subsidence over areas of ~100 km with millimeter-scale vertical resolution. Light detection and ranging (LiDAR) interferometers can measure the velocity or displacement fields of vibrating bodies. In interference light microscopy, vertical resolution can reach below 1 nm, while the Laser-Interferometer Gravitational-Wave Observatory (LIGO) interferometer, developed to detect gravitational waves, can resolve displacements on the order of 10^{-19} m.

Confocal laser scanning microscopy (CLSM)

A focused laser beam scans across the sample to construct high-resolution images from emitted or scattered light. A pinhole aperture rejects out-of-focus light, allowing each scan to capture an image at a specific focal plane, a process known as optical sectioning. By scanning multiple focal planes, the microscope builds a three-dimensional image of the sample.

Cavity ring-down spectroscopy (CRDS)

Conventional light absorption spectroscopy requires long path lengths to detect weak absorption because light intensity decays exponentially with distance.[8] In CRDS, a laser pulse is trapped between highly reflective mirrors, bouncing thousands of times within the cavity. The rate at which the light intensity decays reveals the absorption characteristics of the medium, enabling ultra-sensitive detection of trace gases or liquids.

Atomic force microscopy (AFM)

A sharp tip mounted on a flexible cantilever bends in response to atomic-scale forces, while a laser beam monitors its deflection. Piezoelectric actuators control both the tip and the sample stage, scanning the specimen surface to generate an interaction map with nanometer-scale resolution. AFM can also be used to manipulate matter and construct nanostructures with atomic-level precision.

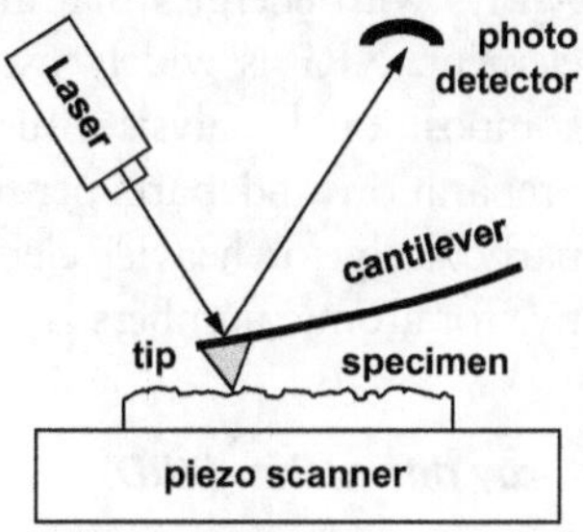

Scanning tunneling microscopy (STM)

Similar in concept to the AFM, the STM also relies on piezoelectric actuators for precise positioning and dynamic control. However, in this case, a small bias voltage is applied between a sharp conductive tip and the electrically conductive or semiconductive sample. This generates a tunneling current across the few-ångström gap between the tip and the surface. The tip scans the surface under careful dynamic control, maintaining a constant tunneling current by adjusting its height, thereby producing an electronic topography map of the surface with atomic resolution.

Electron microscopy: SEM and EDS

A beam of high-energy electrons produced by an electron gun is focused onto a specimen. Electromagnetic lenses scan the beam across the surface by controlling its trajectory via the Lorentz force. The interaction of the beam with the specimen generates different signals: secondary electrons provide high-resolution topographic images in scanning electron microscopy (SEM); backscattered electrons, reflected from deeper layers, are sensitive to atomic number; and characteristic X-rays, emitted when inner-shell electrons are dislodged, enable elemental analysis through energy-dispersive X-ray spectroscopy (EDS).

Transmission electron microscopy (TEM)

In this case, electrons pass through ultrathin specimens (typically <100 nm thick). The transmitted electrons are detected to form two-dimensional images of internal structures, revealing crystallographic orientations, lattice defects and nanostructures with atomic-scale resolution.

X-ray fluorescence (XRF)

While EDS uses electrons to excite the specimen, XRF uses high-energy X-rays. The ejection of inner-shell electrons produces secondary X-rays with energies that are specific to each element. XRF is widely used for rapid, bulk compositional analysis with minimal specimen preparation and parts-per-million sensitivity, particularly for heavier elements (most effective for atomic numbers $Z > 15$).

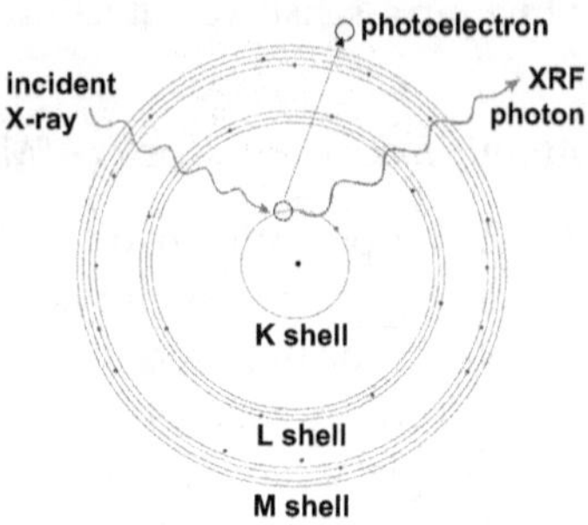

X-ray diffraction (XRD)

When a monochromatic X-ray beam strikes a crystalline material, it is scattered by the electron clouds of atoms in successive crystal planes. Constructive interference occurs when the path difference satisfies Bragg's law, $\lambda = 2d \sin\theta$, where d is the spacing between crystal planes and θ is the scattering angle. The resulting interference peaks, observed at different angles, can be analyzed to determine the crystallographic structure and detect structural changes.

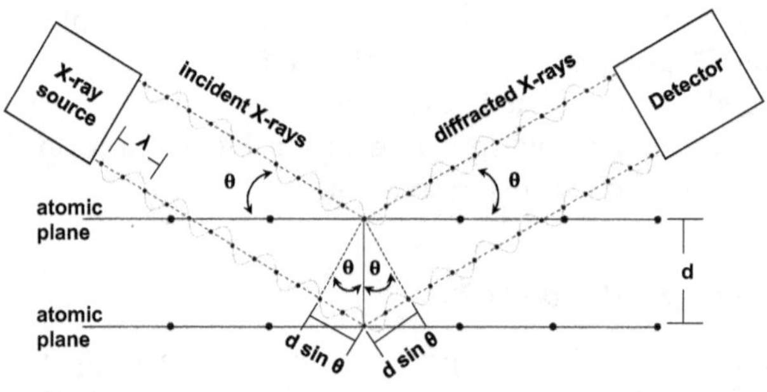

X-ray computed tomography (CT)

X-ray attenuation depends on a material's density and composition. As X-rays pass through an object, their intensity decreases according to the cumulative energy absorbed along the path. A tomograph measures transmitted X-rays from multiple directions across the object. These measurements, called projections, are then processed using mathematical inversion to reconstruct the object's internal structure.

Electrophoretic light scattering (ELS)

Charged colloids in suspension migrate when an electric field is applied. The migration speed depends on particle size, charge and the properties of the surrounding fluid. Laser Doppler electrophoresis measures frequency shifts in scattered light to determine particle mobility and zeta potential. The applied electric field can be oscillated to refine measurements, and complementary titrations may be performed to investigate the effects of pH and ionic concentration on particle charge.

Mass spectrometry (MS)

A solid, liquid or gaseous specimen is ionized, typically by electron bombardment. The resulting ions are then accelerated in an electric field and separated as they travel through a magnetic field, with the degree of separation depending on their mass-to-charge ratio (m / z). When the ions reach the detector, they generate currents at different positions, which are used to produce a mass spectrum plotting intensity versus m / z. Each peak in the spectrum indicates the relative abundance of a specific ion species.

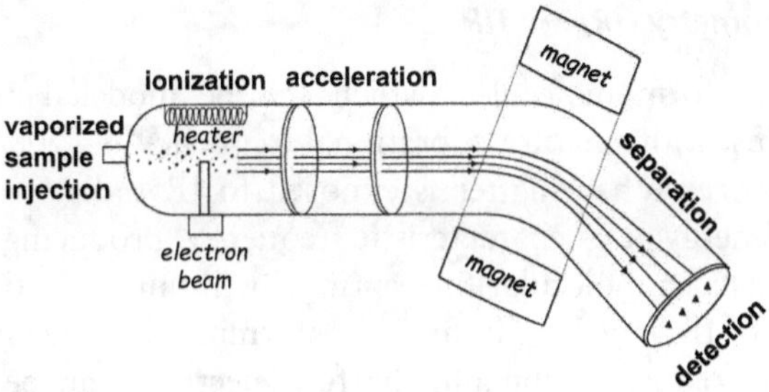

Inductively coupled plasma mass spectrometry (ICP-MS)

A plasma is a gas containing a significant number of ions, making it electrically conductive and responsive to magnetic and electric fields. ICP-MS combines a high-temperature plasma source, which atomizes and ionizes elements in a liquid sample, with a mass spectrometer that separates the resulting ions according to their mass-to-charge ratio m / z. This technique can detect trace elements at concentrations as low as parts per trillion.

Gas chromatography (GC)

Volatile compounds are injected into a column together with a carrier gas. The column may be packed with materials such as diatomaceous earth or silica gel, or it may consist of a capillary tube coated with a stationary phase. As the compounds travel through the column, they interact with the stationary medium and experience different retention times, leading to separation. The individual compounds are then detected at the column outlet using techniques such as flame ionization detection (FID) or gas chromatography-mass spectrometry (GC-MS).

Ion chromatography (IC)

Conceptually similar to GC, IC separates ions in an aqueous solution based on their interactions with a stationary ion-exchange resin. Differences in these interactions result in varying retention times, causing ions to separate as they travel through the column. The eluted species are then detected at the column outlet by monitoring changes in electrical conductivity.

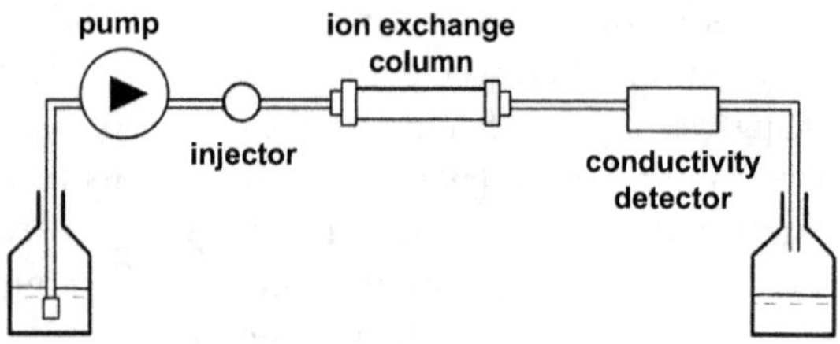

Infrared spectrometry (IR) → FTIR

Atoms bond to form molecules, which can be modeled as mass-m and spring-k systems, with a natural vibrational frequency $\omega = \sqrt{k / m}$ typically in the infrared range. When matter is exposed to IR radiation, each type of bond absorbs energy at its characteristic frequency, producing an absorption pattern that acts as a molecular fingerprint. By illuminating the sample with a broad range of IR wavelengths and transforming the measured absorption signal into the frequency domain, the full spectrum can be reconstructed simultaneously. Fourier-transform infrared spectroscopy (FTIR) enables rapid identification of chemical bonds in solids, liquids and gases.

Nuclear magnetic resonance (NMR)

In diamagnetic materials, such as most organic molecules and water, all electrons are paired.[9] Nevertheless, their nuclear spins can align in a static magnetic field B_0, where they precess at a characteristic Larmor frequency, similar to spinning tops. A transverse magnetic field oscillating at this frequency resonates with the processing spins, effectively tipping them away from the B_0 axis. After the transverse field is turned off, the spins gradually realign along B_0: the longitudinal magnetization recovers with the characteristic time T_1, while the transverse magnetization decays with the characteristic time T_2.

Magnetic resonance imaging (MRI)

The Larmor frequency increases linearly with the static magnetic field B_0. By applying a spatially varying field $B_0(x,y)$, the resonance frequency becomes position-dependent, allowing MRI to reconstruct images that map the local T_1 and T_2 relaxation properties across the (x,y) plane.

Final reflections

We have reviewed a small selection of characterization tools. There are countless other fascinating concepts and instruments for experimental research, from nanoindenters to nuclear accelerators and the Hubble Space Telescope! These devices have revolutionized our ability to observe the physical world, revealing structures, processes and patterns once unimaginable.

Every advanced device introduces layers of processes and sensors beyond our direct perception, with data processing grounded in hypothesized models and assumptions. Most of us use these devices as black boxes, yet progress often depends on understanding their inner workings. In all cases, calibration remains fundamental, for it transforms the device into a comparator between the unknown and the reference we accept as known.

A defining feature of modern instrumentation is the ingenious combination of physical principles. Increasingly, these systems are further augmented by extensive databases and artificial intelligence. Indeed, today's technology is truly remarkable ... almost magical!

Explore a broad range of devices related to your field and beyond. Seek to understand their operating principles, capabilities and limits of resolution. Do not hesitate to adopt new tools or modify existing ones to assemble your own experimental toolbox.

Technology alone is not enough, and sophisticated instruments should never become ends in themselves. The true power of technology in experimental research emerges only when it is guided by a clear purpose, grounded in sound physical understanding and driven by ingenuity and curiosity. Whether using low-tech or high-tech approaches, the path to discovery always begins with a well-defined and meaningful question.[10]

Exercises

1 Visit the most advanced characterization laboratories available at your institution. Identify key instruments and understand their underlying physical principles, main capabilities, operational range and resolution. Reflect on how these tools could be applied to address critical research questions in your field.

2 Study the major diagnostic imaging methods used in medicine: X-ray radiography, CT scans, PET scans, MRI and ultrasound. For each technique, describe the governing physical principle, the material or biological parameters it detects, its spatial and temporal resolution and implementation strategies (e.g., tracers or contrast agents). Include a brief discussion of safety considerations.

3 If your interests lie in materials science, consider methods such as confocal microscopy, thermogravimetric analysis (TGA), goniometry and tensiometry for interfacial properties, nanoindentation, rheometry and the

Brunauer-Emmett-Teller (BET) method for specific surface area determination by gas adsorption. For each technique, explain the physical principle, the parameters it measures and its resolution. Discuss how these instruments could be used to address key questions in your area of research.

4 DNA sequencing and CRISPR genome-editing technologies have transformed biological research and earned multiple Nobel Prizes over the past four decades. Examine their experimental protocols, underlying principles and the fundamental mechanisms of the tools involved.

5 The 2017 Nobel Prize in Physics recognized LIGO. Explore its experimental design in detail – laser power, interferometer arm length, mirror stability and data analysis strategies – that together enable its extraordinary measurement precision.

6 A CHNSO analyzer determines the elemental composition of organic materials: carbon (C), hydrogen (H), nitrogen (N), sulfur (S) and oxygen (O). Study its operation in detail (Hint: combustion in excess oxygen complemented with GC).

Notes

1 C. Bernard (1813–1878), French physiologist.

2 See essays on *Introduction to Devices and Sensors* and *Enduring Simple Machines: From Archimedes to Galileo and Pascal (Section 4).*

3 See essays on *Sensing the Experiment: Scientific Sensors, Electronics: Fundamentals and Applications*; *Waves: Carriers of Energy and Information*; *Optics: Light and Photography (Section 4).*

4 Several sketches in this essay are based on images sourced from the internet.

5 See essay on *Signal Processing and System Analysis (Section 5).*

6 See essay on *Inversion: Inferring Causes from Effects (Section 5).*

7 See essay on *Boundaries: Framing Experiments in Space and Time (Section 4).*

8 See essay on *Waves: Carriers of Energy and Information (Section 4).*

9 See essay on *Electronics: Fundamentals and Applications (Section 4).*

10 See essay on *A Great Question: Where Good Science Begins (Section 1)* and *Creativity and Creative Individuals (Section 2).*

Essay 38

SAFE LABORATORY OPERATION

Discipline sustains discovery

Safety does not happen by accident[1]

Safety is the prerequisite for every laboratory operation. It is never to be taken lightly or compromised by shortcuts. By fostering a culture of care, responsibility and constant attention, safety becomes second nature.

Hazard awareness

Before starting any experiment, fully understand the physical, chemical and biological hazards involved: flammable or explosive substances, high voltage, lasers, ionizing radiation, toxic fumes, reactive chemicals, or extreme pressure, temperature, force or sound. Minimize exposure, review all Material Safety Data Sheets (MSDS) and operate behind protective shields when necessary.

Consider spatial and infrastructural constraints: avoid conducting potentially explosive experiments near flammable materials, account for the needs of others in shared spaces and verify that existing power supplies and utilities are suitable.

Establish emergency procedures and know the locations of exits, fire extinguishers, safety showers and first aid kits. Post emergency contacts along with building and room numbers in accessible locations. Ensure a usable phone or reliable connectivity is available, as laboratories are often in areas with poor reception.

DOI: 10.4324/9781003727521-42

Golden rules

Though their origins will remain anonymous, each of the following guidelines carries the memory of someone who learned the hard way!

- *Dress appropriately.* No open-toed shoes, shorts, skirts, loose clothing, earbuds or dangling accessories. Tie back long hair. Wear required personal protective equipment, including safety goggles, gloves and lab coats.
- *No food or drink.* Eating, drinking or storing food in the lab is strictly prohibited.
- *Label all containers clearly.* Never trust memory or assumptions. Clearly indicate contents, concentration, hazards, date and your name. Yes, label water too!
- *Know your limits.* Always use a buddy system, especially when handling hazardous materials or equipment. Never work alone if you have medical conditions such as a tendency to faint or experience sudden allergic reactions. Emergencies escalate quickly when no one is there to help.
- *Stay alert.* Observe your surroundings. If something seems wrong, speak up or act immediately – your attention could prevent an accident.
- *Dispose of waste properly.* Use designated containers. Never mix incompatible chemicals or discard sharps improperly.

Handle with care

All users are custodians of the lab and its equipment:

- *Inspect and maintain equipment.* Ensure devices are in good condition. Address any potential issues before they become safety hazards. Record failures and repairs in a log accessible to all lab members.
- *Understand your instruments.* Read manuals and consult experienced users. Systems are often designed to be foolproof; yet human negligence can easily exceed the designer's imagination. Do not leave equipment unattended unless it has been explicitly designed and prepared for long-term operation. In such cases, implement appropriate safeguards (e.g., pressure relief, interlocks) and inspect the setup periodically. Use resources responsibly and treat all devices with care: equipment may fail, but misuse remains the leading cause. Carelessness is unacceptable!
- *Protect computers.* Safeguard data, prevent unauthorized access (never share passwords), avoid unverified software and report malfunctions promptly.

- *Respect ongoing experiments*. Never remove parts, tools or components from someone else's setup without explicit permission. This can ruin weeks of work and undermine lab operations.
- *Do not move tools or equipment between labs*. Unauthorized "traveling" items often disappear, are unavailable when needed and create hazards if maintenance is missed.
- *Leave the lab cleaner than you found it*. Return tools and equipment to their proper places, dispose of trash safely and clean the workspace. Order and neatness improve both safety and efficiency.

Safety in focus

Safety is the *sine qua non* of experimentation. Lapses often arise from time pressure, overconfidence or routine familiarity. Regularly review safety guidelines and conduct drills. Design experiments to prevent accidents before they occur. Much like the culinary principle *mise en place*, where ingredients and tools are organized before cooking, safe experimentation begins with careful preparation: verifying materials, equipment, procedures and contingencies. Disciplined attention and genuine care quietly make meaningful experimentation possible. *Safety begins long before the first experiment*!

Exercises

1 Compute the energy in a 100 cm^3 pressure vessel filled with air at 10 MPa. Compare it with the gravitational potential energy of a person standing 1 m above the ground. What does this comparison imply about the potential hazards of pressure vessels and the importance of safe handling?
2 Prepare a laboratory safety evaluation checklist for your field of interest. Base your assessment on applicable regulations, codes, and best practices. Include criteria for personnel safety, equipment, hazardous materials, emergency preparedness, and documentation.
3 Assess your laboratory's safety (including equipment, materials, and operations) using the checklist developed in the previous exercise. Identify the main hazards, verify emergency procedures and posted contact information; inspect safety equipment and signage. Recommend measures to improve safety and reduce risk.

Note

1 Common saying in ergonomics.

SECTION 5

Analytical and numerical tools. The role of analysis

Essay 39

INTRODUCTION TO DATA PROCESSING AND MODELING

In theory, theory and practice are the same. In practice, they are not[1]

Researchers gradually discover their comfort zone: some gravitate toward equations and simulations, while others thrive in the laboratory. But comfort can quietly turn into constraint, and relying on a single mode of inquiry can limit both insight and impact.

Experimentalists, in particular, may miss valuable opportunities by avoiding even basic analytical considerations. Consider the deceptively easy task of measuring heat transfer in a small cylindrical specimen: the device may be meticulously assembled and the sensors properly calibrated; yet without considering the governing heat-transfer equations and relevant dimensionless parameters, measurements may miss the transient response, or the actual boundary conditions may diverge from the intended ones. Soon we fall into a familiar trap: being unable to distinguish what we want to measure, what we think we measure and what we actually measured.

A balanced strategy helps avoid these blind spots. Experiments probe the real world within the constraints of instruments, materials and feasible conditions. Simulations provide controlled, reproducible "virtual experiments", though they inevitably embed assumptions. Analyses often yield profound insights and plausible predictions, but only for abstractions of reality. Each approach carries limitations, but together they sharpen, constrain and challenge one another. In particular, analysis and simulation can:

- *Strengthen experimental design*: identify critical variables, define informative parameter ranges and guide sensor selection, placement and boundary-condition testing.

DOI: 10.4324/9781003727521-44

- *Reveal hidden processes*: illuminate internal fields and governing processes that cannot be directly measured.
- *Broaden experimental reach*: provide a framework to upscale results, extrapolate to new conditions and explore regimes beyond physical limits (nanoseconds, kilometers, gigapascals and extreme temperatures).
- *Accelerate exploration*: enable rapid study of parameter combinations that would be prohibitively expensive to test experimentally.
- *Reduce cost and risk*: especially valuable for hazardous, destructive or logistically demanding studies.

This section invites you to integrate analytical and numerical thinking into experimental practice. These introductory essays are not tutorials but encouragement to expand your investigative toolkit. Embracing models and simulations alongside experiments enforces rigor, clarifies assumptions, extends the reach of your findings and amplifies the overall influence of your work.

Exercise

Identify three recent research questions you have worked on. For each one: did you perform a scaling analysis (e.g., using Buckingham's Pi theorem), apply analytical reasoning or run numerical simulations while exploring the problem and planning the experiments? If yes, what did you learn from these efforts? If not, how might they have helped?

Note

1 Although often attributed to Y. Berra (1925-2015), the saying originates in the 19th century.

Essay 40

SIGNAL PROCESSING AND SYSTEM ANALYSIS

Nothing is complex.
Imaginary tried hard-
Phase was all it was[1]

Signal processing seeks to extract meaningful information from data streams, often to infer the behavior of underlying systems. Key historical developments include J. Fourier's trigonometric series (1822), the Nyquist-Shannon sampling theorem (H. Nyquist, 1928; C. Shannon, 1948) and the fast Fourier transform introduced by J. Cooley and J. Tukey (1965), which helped launch the digital signal processing revolution.

This essay highlights the most critical concepts. For clarity, we consider a one-dimensional array of N points $x = \left[x_0, x_1, \ldots x_i, \ldots, x_{N-1}\right]$ sampled at times $t_i = i\Delta t$ with constant sampling interval Δt (Note: the signal may also represent a spatial profile with sampling interval $\Delta \ell$).[2]

LTI systems and elemental signals

Linear time-invariant system

Signal processing typically assumes superposition: the sum of causes produces the sum of their effects, a property that holds for linear systems. Algorithms also implicitly assume that identical inputs applied at different times generate the same response, meaning the system does not change over the observation period. Together, these assumptions define a central premise of classical signal processing: the system is *linear time-invariant* (LTI).

DOI: 10.4324/9781003727521-45

Elemental signals

Several "elemental" signals play an essential role in the analysis of signals and systems in engineering and scientific applications. Two of the most important are the impulse and sinusoidal signals. In digital form:

- The impulse δ_i has zero value everywhere except at the origin, where its value is 1.0; mathematically,

$$\delta_i = 1 \text{ if } i = 0, \text{and } \delta_i = 0 \text{ if } i \neq 0.$$

- A sinusoidal signal is a sine (or cosine) function evaluated at discrete times $t_i = i\,\Delta t$,

$$x_i = \sin(\omega\, t_i + \phi)$$

Recording great signals

Improve the signal at the earliest stage

Start with a well-designed experiment, carefully selected sensors, matched impedances and tightly controlled boundary conditions. Minimize all sources of noise *before* acquiring measurements.[3]

Analog-to-digital (A/D) – Aliasing

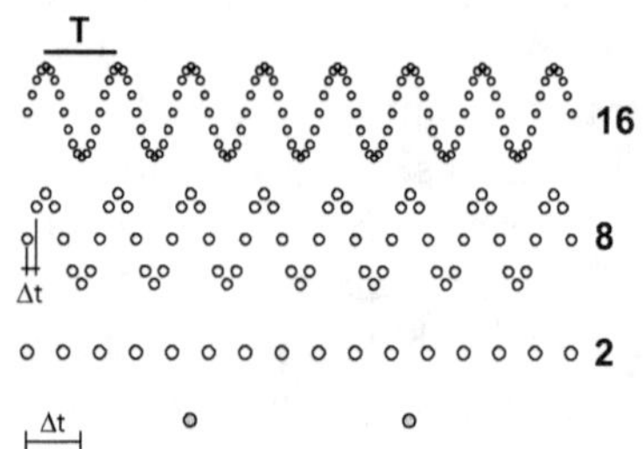

Consider a sinusoid sampled at only two points per period, taken exactly at the zero crossings. When plotted, the digitized values do not resemble a sinusoid but a flat line. Proper sampling, therefore, requires more than two samples per cycle: the sampling interval must satisfy $\Delta t < T/2$ where T is the signal period. More generally, the sampling frequency must exceed the Nyquist criterion for the highest frequency component $f_s > 2f_{\max}$. In practice, 10 or more samples per cycle are often used to capture the waveform accurately. While higher sampling rates improve signal definition, they also increase memory requirements.

Under-sampled signals can appear sinusoidal but at incorrect, "aliased" frequencies. To prevent aliasing, a low-pass antialiasing analogue filter should be applied before the A/D conversion (Note: such filters are standard in digital storage oscilloscopes).

Signal stacking

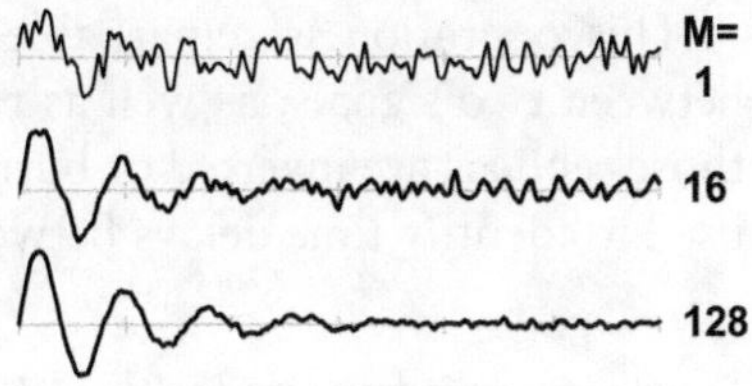

This is a robust strategy to improve the signal-to-noise ratio (SNR). The process involves measuring the signal M times and averaging all M values at each time step i to obtain the stacked value:

$$x_i^{\langle \text{avr} \rangle} = \frac{1}{M} \sum_{k=1}^{M} x_i^{\langle k \rangle}$$

Effective signal stacking requires four key conditions: (1) the source must be repeatable; (2) the system must remain constant during the stacking process, i.e., time-invariant; (3) the trigger used to start each recording must be well synchronized to avoid random time shifts between measurements; and (4) noise must have zero mean value at every t_i.

Signal processing

Finding similarities - Cross-correlation

While our eyes can easily detect similarities between two clear signals, a computer only sees two arrays of numbers. How can it detect similarities? The answer is the cross-correlation algorithm.

For two discrete signals x and y, the cross-correlation is computed by multiplying corresponding elements and summing over all indices. Then, shift the signal x by one step relative to y, and repeat again. The sequence of values for different k-shifts defines the cross-correlation array:

$$\text{cc}_k^{\langle x,y \rangle} = \sum_i x_i \, y_{i+k}$$

Consider the following two examples:

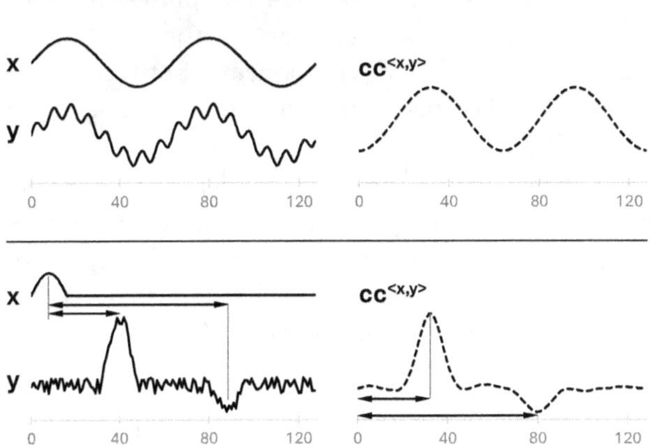

This operation is remarkable! It detects shared sinusoidal components between two signals, as well as replicas of the signal x within y, even when those replicas are inverted or buried in noise. Cross-correlation is also widely used to identify time delays between signals.

Fourier transform (Similarity with sinusoids)

We can "curve fit" any discrete signal with a Fourier series

$$x_i = \sum_{u=-\infty}^{\infty} \left[A_u \cos(\omega_u t_i) + B_u \sin(\omega_u t_i) \right]$$

$$= \sum_{u=-\infty}^{\infty} \left[A_u \cos\left(u \frac{2\pi}{N\Delta t} t_i \right) + B_u \sin\left(u \frac{2\pi}{N\Delta t} t_i \right) \right]$$

The selected frequencies are harmonically related. For $u = 1$, the sinusoid spans the entire signal duration $N\Delta t$, with frequency $\omega_1 = 2\pi / N\Delta t$; other sinusoids have frequencies $\omega_u = u(2\pi / N\Delta t)$. Because the cross-correlation between any pair of harmonically related sines or cosines is zero, these sinusoids form a "basis" that supports the unambiguous projection of the signal onto the series of sinusoids.

There are several equivalent ways to express the sine and cosine components. For a given frequency ω_u:

$$A_u \cos(\omega_u t_i) + B_u \sin(\omega_u t_i) = R_u \sin(\omega_u t_i + \phi_u)$$
$$= \Re\left[(A_u - jB_u) e^{j\omega_u t_i} \right] = \Re\left[X_u e^{j\omega_u t_i} \right].$$

The second form combines the sine and cosine into a single amplitude R_u and phase ϕ_u. The third form uses Euler's identity to express the (A_u, B_u) pair as a "complex" number $X_u = A_u - jB_u$, with amplitude $R_u = |X_u| = \sqrt{A_u^2 + B_u^2}$ and phase $\phi_u = \operatorname{atan}(B_u / A_u)$. We call the A-factors the "real" components and the B-factors the "imaginary" components. But there is nothing imaginary or complex in this compact representation: all values are very real and quite simple indeed!!

The array of coefficients $X = [(A_0, B_0), (A_1, B_1), \ldots]$ *or* $X = [X_0, X_1, \ldots]$ is the Fourier transform of the signal $X = \mathrm{FT}(x)$. Each coefficient measures how similar the signal is to a sinusoid of frequency ω_u, effectively representing the cross-correlation between the signal and that sinusoid. Then, the analysis equation in terms of complex exponentials is:

$$X_u = \sum_{i=0}^{N-1} x_i \, e^{-j\left(u \frac{2\pi}{N} i \right)} \qquad \textit{Analysis – The Fourier Transform FT}$$

Conversely, the signal x can be reconstructed by summing its sinusoidal components:

$$x_i = \sum_{u=0}^{N-1} X_u e^{j\left(u\frac{2\pi}{N}i\right)}$$ *Synthesis – The Inverse Fourier Transform IFT*

These two expressions form a Fourier pair. Several such pairs exist; once a particular analysis equation is chosen, its corresponding synthesis equation must be used to correctly reconstruct the signal.

System characterization with elemental signals

Hit it with an impulse: the impulse response

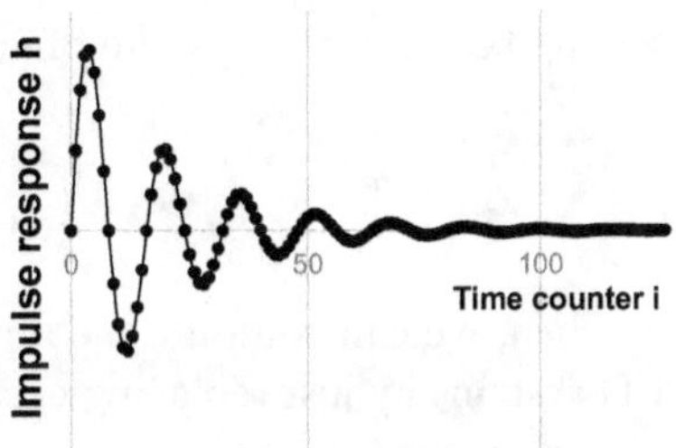

Tap a bell or strike a long, flexible ruler with a short impact, and you will hear or see vibrations as they decay over time. The digitized output signal is the impulse response array h. In practice, the "true impulse" is not physically realizable, and the short physical impact is only an approximation of the mathematical impulse described earlier. The actual impulse response h is computed mathematically, as discussed below.

Excite it with a sinusoid: the system response

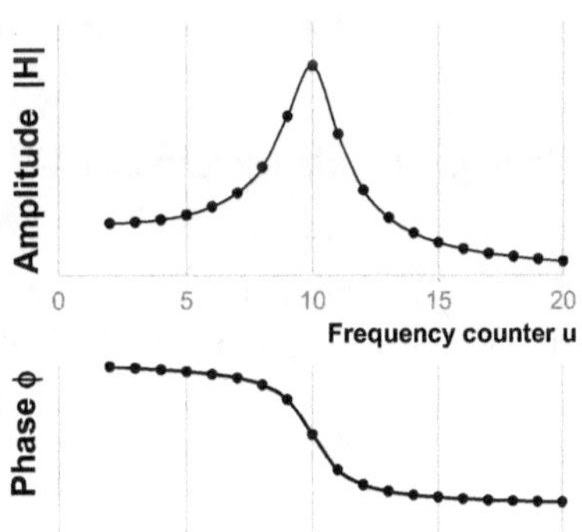

The response of a LTI system to a sinusoidal input is another sinusoid of the same frequency ω_u, of amplitude A_u and phase-shifted $\Delta\phi_u$. (Note: sinusoids are called eigenfunctions of LTI systems). By applying sinusoidal excitations at different frequencies ω_u, we can measure the resulting amplitudes A_u and phase shifts $\Delta\phi_u$. The collection of these measurements forms the system frequency response array H.

How are they related?

Both the impulse response h and the frequency response H contain the same information and completely characterize a linear time-invariant system. Consequently, they must be mathematically equivalent; indeed, the frequency response is the Fourier transform of the impulse response $H = \mathrm{FT}(h)$. Conversely, we can obtain the impulse response by taking the inverse Fourier transform of the frequency response: $h = \mathrm{IFT}(H)$.

System response to any input - Convolution

Once the system has been characterized with elemental signals, we can compute its output y for any input x.

Using the impulse response

It may be obvious but is still very valuable to highlight that a signal x can be represented as a sequence of scaled and time-shifted impulses δ_{i-k}, where the scaling factor at the discrete time t_k is the kth value x_k of the signal:

$$x_i = \sum_k x_k \, \delta_{i-k}$$

Then, we can compute the output y using the superposition principle for LTI systems by just replacing each time-shifted impulse δ_{i-k} by a time-shifted impulse response h_{i-k}:

$$y_i = \sum_k x_k \, h_{i-k}$$

In words: the sum of scaled and time-shifted impulses produces the sum of equally scaled and time-shifted impulse responses.

Using the frequency response

The Fourier transform of an input signal $X = FT(x)$ quantifies how much energy the signal contains at each sinusoidal frequency. On the other hand, the frequency response H describes how an input sinusoid is modified by the system in both amplitude and phase. Then, the output at a given frequency ω_u is

$$Y_u = H_u \, X_u$$

The array Y of Y_u values represents the Fourier transform of the output. Then, the sought output signal y is the inverse Fourier transform: $y = \mathrm{IFT}(Y)$. Be aware that some Fourier pairs do not preserve energy in the convolution.

In time or in frequency?

Computing convolutions in the frequency domain using the Cooley-Tukey fast Fourier transform (FFT) algorithm is typically much faster than performing them in the time domain. However, trade-offs are unavoidable: while the input signal x lasts $t = N\Delta t$, the Fourier transform implicitly assumes that the signal is periodic, repeating every $N\Delta t$. This can lead to circular convolution, where energy in the computed output y appears before the actual excitation in the input x, originating from the "previous cycle" in the theoretical Fourier representation.

In closing: great tool, but little devils everywhere

Signal processing is an essential tool in the experimentalist's toolbox. Its algorithms are ingenious, rarely complicated and remarkably powerful. These qualities make signal processing both appealing and potentially hazardous. Use it wisely: understand the underlying assumptions and apply the methods judiciously. Pitfalls are everywhere; remain alert.

Many operations described above rely on summations, which inherently assume linearity and time invariance. If the system is nonlinear or time-varying, discrete computations will still produce an answer, but it may be misleading.

The key takeaway is clear: always start with a well-designed experiment. Once you have reliable data, apply signal processing thoughtfully. Done properly, it can deliver extraordinary insight; done carelessly, it can mislead more than it informs.

Exercises

1 *Aliasing*. Spin a wheel (e.g., a bicycle wheel, toy wheel, or fan) and record it using a digital camera. Note: a typical frame rate of 30 frames per second corresponds to a sampling interval of $\Delta t \approx 33\,\text{ms}$. Play the video back in slow motion. You may observe that the wheel appears to rotate more slowly than it actually does, or it may even seem to spin backward while physically moving forward. Discuss in detail why this happens, relating your observations to aliasing and discrete-time sampling.

2 *Cross-Correlation*. Compute, by hand, the cross-correlation between the following two signals:

$$x = [0\ 0\ 1\ 2\ 1\ 0\ 0\ 0\ 0\ 0\ 0\ 0\ 0\ 0\ 0\ 0\ 0\ 0\ 0\ 0\ 0\ 0\ 0\ 0\ 0]$$

$$y = [1\ 1\ 1\ 2\ 3\ 2\ 1\ 1\ 1\ 1\ 1\ 1\ 1\ 1\ 1\ 1\ 1\ 1\ 0\ {-1}\ 0\ 1\ 1\ 1\ 1]$$

Show the steps as you shift one signal relative to the other and compute the summation at each shift.

3 *Convolution.* How much will your car bounce $\boldsymbol{c}$ when it travels over a speed bump? Compute the convolution of the speed bump $\boldsymbol{b}$ with the car's impulse response $\boldsymbol{h}$ (Note: The speed bump is defined as height vs. distance, but since the car moves with constant velocity, we converted it to height vs. time).

$$\boldsymbol{b} = \begin{bmatrix} 0 & 0 & 0 & 0 & 1 & 2 & 3 & 4 & 4 & 4 & 4 & 3 & 2 & 1 & 0 & 0 & 0 & 0 & 0 & 0 & 0 \end{bmatrix}$$

$$\boldsymbol{h} = \begin{bmatrix} .5 & .8 & 1 & .2 & 0 & -.1 & 0 & 0 & 0 & 0 & 0 & 0 & 0 & 0 & 0 & 0 & 0 & 0 & 0 & 0 & 0 \end{bmatrix}$$

4 *Frequency response from a broadband input.* You do not need to perform a frequency sweep to obtain the values H_u in the frequency response $\boldsymbol{H}$. A more efficient method is: (a) Excite the system with a broadband input $\boldsymbol{x}$ and measure the output $\boldsymbol{y}$. (b) Compute the Fourier transforms $\boldsymbol{X}$ and $\boldsymbol{Y}$. (c) Compute the frequency response at each frequency ω_u: $H_u = Y_u / X_u$. (d) Assemble all values to obtain the full frequency response $\boldsymbol{H}$. (e) If needed, compute the impulse response: $\boldsymbol{h} = \text{IFT}(\boldsymbol{H})$. Apply this procedure to a system of your choice.

Notes

1 A haiku for signal processing. This traditional Japanese poetic form has a characteristic structure based on 5-7-5 syllables.
2 This essay is extracted from J.C. Santamarina and D. Fratta (2005). *Discrete Signals and Inverse Problems.* John Wiley & Sons.
3 Review essays in Section 4, in particular: *Noise... What a Pain!*.

Essay 41

CORRELATION OR CAUSATION?

Cum hoc ergo propter hoc
(with this, therefore because of this)[1]

Causation is inferred, not observed. Observational data may reveal associations, but true causal inference requires counterfactual reasoning by testing the same system under alternate conditions or by exploiting natural experiments that expose different groups to different circumstances. Regression analyses must also include the important confounding variables that influence both the factor of interest and the outcome. Yet, despite our best efforts, causal reasoning can be misled by several common pitfalls.

Correlation does not imply causality (false positives)

Two variables may correlate without either one causing the other: a *confounding variable* may drive both. Colds are more frequent in winter, not because cold air causes illness, but because seasonal conditions – such as indoor crowding and longer viral survival at low temperatures – influence both exposure and infection rates. We may also be fooled by *spurious correlations* or strange statistical coincidences, such as the parallel rise in transistor counts in computer chips and in global sea level.

Some correlations do reflect genuine causal links, but establishing them requires multiple lines of evidence and careful reasoning. For example, the causal effect of increasing carbon dioxide on global warming was established only after extensive physical theory, controlled experiments and detailed observations all supported the same conclusion. In all cases, we must resist

DOI: 10.4324/9781003727521-46

the temptation to read cause where there is only a pattern: treat correlations as clues, never as conclusions.

Lack of correlation does not imply lack of causality (false negatives)

Causal relationships can be hidden when we observe only part of a system: leaving out important dimensions can obscure real patterns. Consider a three-dimensional relation in which z depends on both x and y. A plot of z against x alone may look like random scatter, and the same can occur when z is plotted against y. Yet, in the full three-dimensional space, the pattern becomes clear. What appears as noise in lower-dimensional projections is the shadow of a genuine causal relationship in the higher-dimensional space, a phenomenon often referred to as *projection noise*.

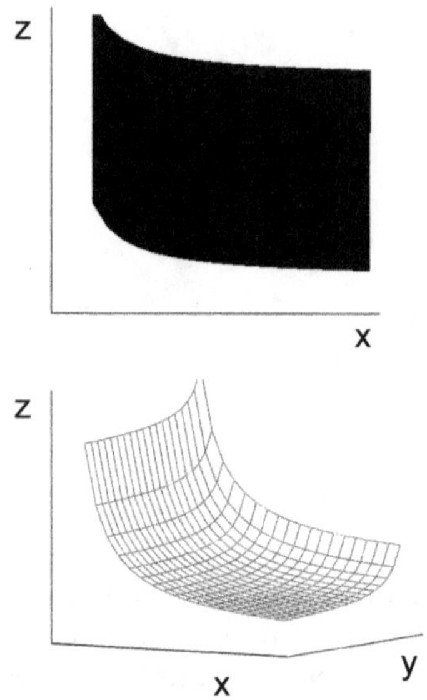

Something similar happened in early studies of smoking and lung cancer. Early analyses did not reveal a clear pattern until factors such as age, gender and the long delay between exposure and disease were taken into account. Once these were included, the causal link became undeniable. Careful model specification and thoughtful selection of covariates are essential when searching for causal structure.

Causality... in what direction? (misdirected arrows)

Even when a causal relationship exists, its direction is not always obvious. Correlations can reflect reverse causation or even feedback loops. Gun ownership and violent crime, for example, can drive each other: rising crime may prompt more gun ownership, while increased gun ownership may contribute to more crime. A similar ambiguity appears in the claim that "successful researchers are better communicators": research experience may sharpen communication skills, strong communication may enhance funding success, or the two may reinforce each other. Temporal order helps – causes must precede effects – but it does not always settle the issue: traffic jams before dawn do not cause the sun to rise.

Do not let the data fool you

Causal inference can fail when we see causation where none exists, overlook it where it does, or misjudge its direction. These failures are not flaws in nature but flaws in interpretation. Patterns can guide our attention, but they cannot, on their own, establish cause. Only a coherent theory, tested through carefully designed experiments and rigorous analysis, can turn patterns into genuine understanding. While this purposeful approach cannot guarantee causality, it makes inferences far more credible.[2]

Exercises

Identify examples from your field that illustrate: (a) correlation without causation – variables appear related but one does not cause the other; (b) hidden causality – a causal link exists but is not immediately apparent; and (c) unclear causal direction – it is uncertain which variable causes the other.

Notes

1 This is the traditional Latin term for the fallacy of assuming that correlation implies causation. Just because two events occur together does not mean that one causes the other.
2 D. Hume (1711–1776), Scottish philosopher, historian and essayist.

Essay 42

MODELS AND FUNCTIONS

Balancing elegance and reality

The map is not the territory...[1]

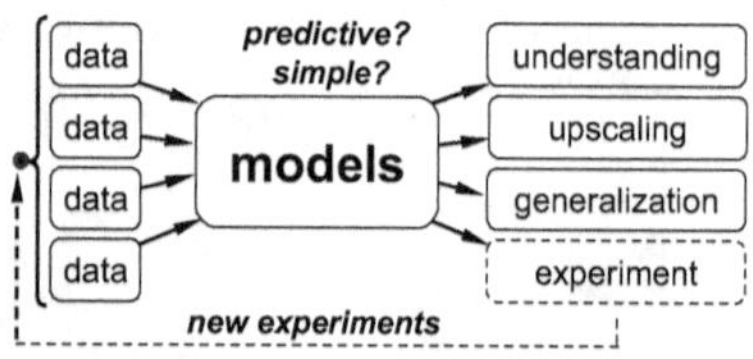

We advance understanding by developing models that organize data and observations into coherent frameworks. The process involves iterative loops. Models shape our perception of reality because we interpret the world through the frameworks we have come to trust. At the same time, they guide the design of new experiments, generating data that can confirm, refine or overturn the frameworks that motivated them. Trade-offs are inevitable: simpler, clearer models often produce less precise predictions, since clarity requires simplification. Ultimately, the purpose of a model is not to deliver absolute truth but to provide insight and generate useful generalizations across the spatial and temporal scales relevant to the system under study. With this perspective in mind, let's look more closely at the role of models in experimental research.

The role of models[2]

Models are abstractions that deliberately omit what is judged irrelevant, suppress complexities and idealize certain features to make analysis possible. These abstractions are not flaws as long as they do not distort the causal connections in the system under study. Nor do models always require radical simplification: modern science increasingly relies on detailed representations, particularly in computational simulations. Yet even the most detailed

DOI: 10.4324/9781003727521-47

numerical models operate within the assumptions required to transform reality into a digital representation.

The central challenge is deciding what to neglect. A successful model navigates the trade-off between tractability and fidelity: enough simplification to enable investigation, but not so much that it misleads. This tension also reflects other implications: highly predictive models may be opaque, whereas clearer ones may sacrifice predictive power.

Improper simplifications can obscure essential behavior or introduce systematic bias, and relevance is rarely known a priori. Model development is therefore a disciplined, iterative process, constrained by theory, tested against data and evaluated alongside alternative models. When predictions fail to align with observations, assumptions must be revisited and understanding revised. Without measurement and empirical confrontation, claims remain speculative.[3]

Deep understanding rarely emerges from exploring a single model. Partially valid models can provide complementary perspectives, forming a patchwork that collectively illuminates phenomena. When models are well chosen and empirically anchored, they reveal genuine insight; conversely, elegance without empirical grounding can be profoundly misleading. Ultimately, a model's value lies not in superficial resemblance to reality, but in its ability to explain, predict and withstand competition from rival representations.

Criteria for selecting promising mathematical functions

Selecting an appropriate functional form for our models is not always straightforward, especially when experimental data span only a narrow range of the control variable, allowing many functions to fit the observations. The following *physics-inspired and data-driven* criteria help identify the most promising options:

1. Physical plausibility and asymptotic correctness: the function must reflect physically justified trends and exhibit correct asymptotic behavior beyond the measured data range.
2. Consistency with physical constraints: the function must satisfy conservation laws (mass, energy, momentum, charge).[4]
3. Structural simplicity with sufficient expressiveness: the mathematical form should be as transparent as possible while still capturing the essential physical effects (parsimony without loss of relevance).[5]

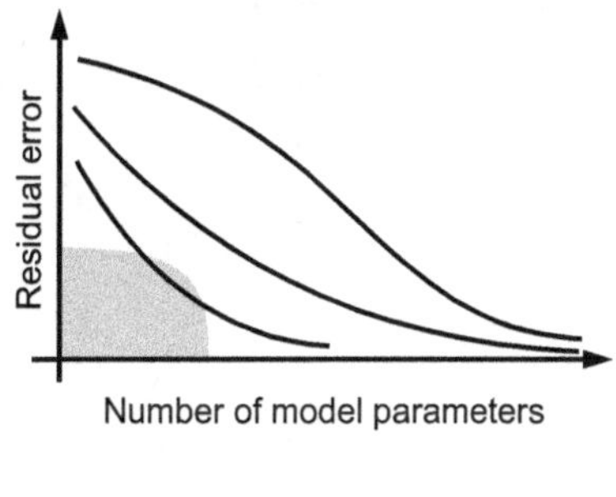

4. Balance between goodness-of-fit and model complexity: the function should fit the data well while avoiding overfitting by limiting the number of free parameters relative to data volume

and noise (Ockham's principle).[4] As A.N. Whitehead cautioned, *seek simplicity, but distrust it.*

5 Smoothness and computational robustness: the function should be smooth and differentiable to ensure stable numerical computation, reliable parameter estimation and efficient evaluation of derivatives.
6 Parameter identifiability: model parameters must be uniquely recoverable from the available data.
7 Stability under perturbations: the function should not produce large changes in predictions due to small variations in input variables or experimental noise.
8 Dimensional consistency: all terms in the function must respect correct physical units or nondimensional structure.[6]

Common mathematical functions in engineering and science

Models rely on mathematical functions. A function by itself has no intrinsic meaning; its significance arises only when we identify the parameters that govern the underlying physical processes. Familiarity with common functions allows us to recognize patterns and choose representations that are both physically meaningful and computationally robust. The functions listed below are frequently encountered in engineering and the sciences.

Function	*Equation*	*Typical Applications/Examples*
Linear	$y = bx$	Ohm $v = Ri$, Coulomb $T = \mu N$, Darcy $v = ki$, Hooke $\sigma = E\varepsilon$, Winkler $F = kd$, Newton $F = ma$, Fourier $q_T = k_T i_T$, Fick's diffusion $J = Di_c$, thermal expansion $\varepsilon = \alpha \Delta T$
Polynomial	$f(x) = a_0 + a_1 x + a_2 x^2 + \cdots$	Approximates complicated trends. Higher-order improves fit but reduces predictability (Ockham's principle)
Exponential	$f(x) = Ae^{kx}$	Attenuation in space/time (skin depth), transient responses (NMR), population growth
Logarithmic	$f(x) = \log(x)$	Order-of-magnitude scaling: pH, decibels
Power	$f(x) = kx^{\alpha}$	Empirical correlations: stress-dependent stiffness, fractals, scale invariance, Kleiber's law

(*Continued*)

Function	*Equation*	*Typical Applications/Examples*
Hyperbolic	$y = a / x$	pV = constant, capacitive impedance $\lvert z_C \rvert = \frac{1}{2\pi C} \times \frac{1}{f}$
	$y = x / (a + bx)$	Nonlinear behavior with asymptotes: For $x \to 0: dy / dx = a^{-1}$. For $x \to \infty: y_{max} = b^{-1}$
Sinusoidal and complex exponential	sin, cos $e^{i\theta} = \cos\theta + i\sin\theta$ (Euler)	Harmonic motion, eigenfunctions of linear time-invariant systems, Fourier decomposition
Hyperbolic sinusoids	$\sinh(x), \cosh(x)$	Catenary curve, linear differential equations with exponential solutions
Gaussian	$f(x) = \frac{1}{\sqrt{2\pi\sigma^2}} e^{-(x-\mu)^2/2\sigma^2}$	Measurement errors, central limit theorem
Sigmoid/ logistic	$f(x) = 1/(1 + e^{-x})$	Population dynamics, neural networks

Notice how widely used linear functions are. Beyond their clarity, linearity enables superposition: *the response to multiple inputs equals the sum of individual responses*. Furthermore, any function can be approximated locally by a linear function using a Taylor expansion.

Closing thoughts

There is an underlying tension between models and physical reality. Analytical models help us identify the most important variables and highlight their interactions through elegant expressions. In contrast, fine-grained numerical models can capture complex processes and accommodate flexible boundary conditions, but they are often opaque and difficult to interpret. Advancing understanding through iterative model development requires balancing clarity with realism.

Exercises

1 Examine a historical case in your field. Analyze the synergism between experimental research and iterative model development in advancing understanding.

2 Collect a dataset from a lab experiment. Fit both a linear model and a higher-order polynomial model. Compare and discuss model parsimony, accuracy within the dataset and potential for misleading extrapolations.

3 Identify commonly used models and functions in your field. Evaluate their adequacy using the criteria for selecting promising models discussed above. If a model fails one or more criteria, propose alternative formulations or models.

Notes

1 A. Korzybski (1879–1950), Polish-American philosopher and scientist, on the limits of abstractions.
2 Inspired by a discussion of models in P.A. Baran and P. Sweezy (1966), *Monopoly Capital*, and expanded to capture modern concepts in the philosophy of science by N. Cartwright, I. Hacking, R. Giere and E. Winsberg.
3 W. Thomson, aka Lord Kelvin, in Popular Lectures and Addresses vol. 1 (1889) 'Electrical Units of Measurement', delivered 3 May 1883:

> *When you can measure what you are speaking about, and express it in numbers, you know something about it; but when you cannot measure it, when you cannot express it in numbers, your knowledge is of a meagre and unsatisfactory kind; it may be the beginning of knowledge, but you have scarcely, in your thoughts, advanced to the stage of science.*

4 See essay on *Our Physical Universe: Laws and Principles (Section 1).*
5 R. Feynman (1985). *The Character of a Physical Law*. MIT Press.
6 See essay on *The Dimensional and the Dimensionless (Section 3).*

Essay 43

ERROR STATISTICS AND PROPAGATION

He uses statistics as a drunken man uses lamp posts – for support rather than illumination[1]

All measurements carry uncertainty. Therefore, the statistical tools developed over the last five centuries have become essential in the experimentalist's toolkit. While modern experimentation relies heavily on statistics, statistical methods alone cannot compensate for weak physics. Model misspecification, such as omitting relevant variables or assuming incorrect governing processes, introduces fundamental biases that propagate through predictions and can easily overshadow measurement noise. These vulnerabilities emphasize the need for deep physical insight, careful experimentation and a thorough awareness of underlying assumptions before drawing conclusions from data.

We introduced statistical concepts earlier, as part of experimental design.[2] Here, we return to the theme of statistics with an emphasis on data analysis.

Biased data structures – The perils of partial data

The most dangerous errors in experimental studies are not random; they are *structural*. Biased sampling, survivorship distortion and selective reporting can mislead studies even when individual measurements are executed with great care. These represent critical caveats: no amount of analysis can salvage data that fail to represent the underlying population or system.

DOI: 10.4324/9781003727521-48

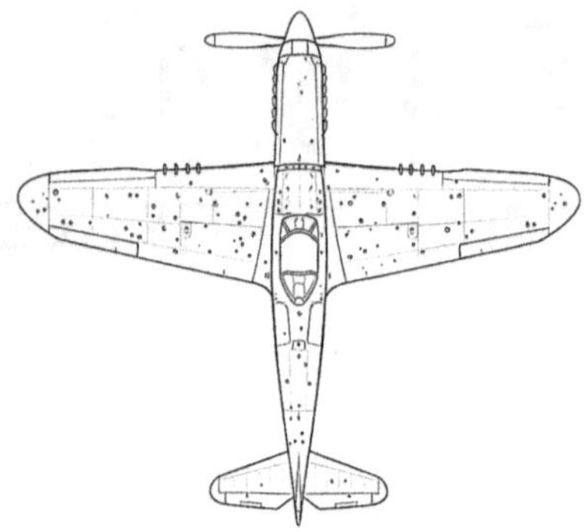

A common selection bias is the emphasis on "survivors" while overlooking "non-survivors". A classic example is the World War II airplane analysis, where initial efforts focused on reinforcing the areas on returning planes with the most bullet holes. A. Wald realized that these were the planes that survived despite being hit in those areas; the true vulnerabilities were the locations *without* bullet holes, because aircrafts struck there never returned.

Underrepresentation of the most affected population is a related bias. In biomedical studies, insufficient recruitment of vulnerable groups and inadequate attention to their specific needs mean that guidelines and risk assessments may reflect incomplete or skewed data, leaving them without optimal care or preventive strategies. A notable case involves uterine fibroids, which disproportionately affect Black women; yet, historically, they have been underrepresented in clinical trials and epidemiological studies.

Publication bias intensifies these distortions: the scientific record favors positive findings, creating a distorted empirical landscape in which careful studies that found "no effect" are less visible. Readers may then conclude that a phenomenon is real when, in fact, the balance of the total evidence points to its absence.

Inadequate controls for confounding factors, survivorship bias and publication reinforcement can lead to – and perpetuate – incorrect conclusions. Consider the case of alcohol consumption and its supposed health benefits. There are confounding factors (e.g., overall healthier lifestyles among moderate drinkers), selective populations (e.g., exclusion of former heavy drinkers who quit, which avoids misclassification but biases results) and a publication tendency to favor the prevailing consensus (i.e., the protective effects of moderate alcohol consumption). When subsequent studies and analyses corrected for these issues, they found no clear health benefits and, in some cases, evidence of harm even at low levels of alcohol consumption.

Distinguishing accuracy and precision: bias and variance

Tightly clustered results can conceal bias. In statistical terms, precision refers to the variance, or the degree to which repeated measurements scatter around a target value, while accuracy refers to bias, the systematic deviation that causes measurements to consistently miss the true value.

A dataset can be highly precise yet inaccurate, leading to misleading conclusions. History provides striking examples. The gravitational constant G

was first measured by H. Cavendish in 1798 using a torsional balance and lead spheres. Since then, numerous measurements have achieved impressive precision, yet reported values have shifted significantly over time, often lying outside the error bars of previous precise measurements (current uncertainty ≈ 22 ppm). Similar historical patterns have been observed for the speed of light and the charge of the electron.[3]

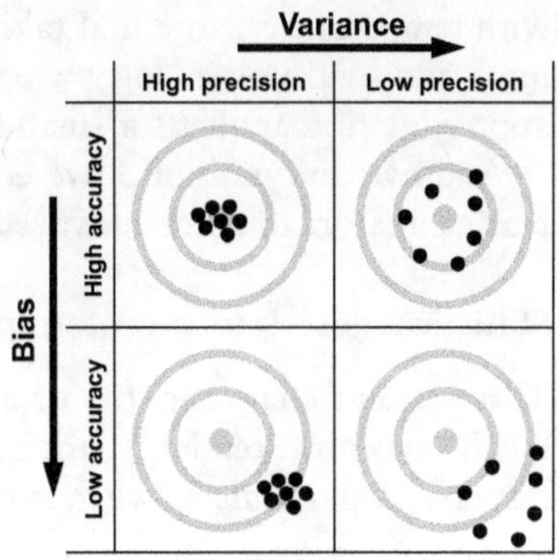

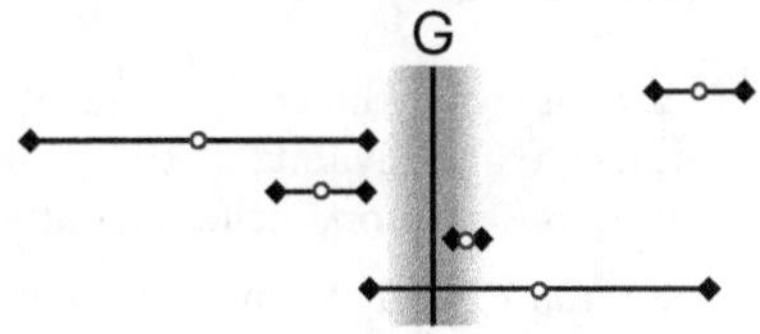

Effective experimental design requires addressing bias and variance from the outset. Precision must match the scale of the problem and remain consistent across repeated measurements. Instrument calibration limits, including resolution, drift, nonlinearity and saturation, define the operational range within which measurements are reliable. These constraints impose hard bounds on achievable accuracy and precision. Ultimately, irreducible noise limits how much information data can reveal.

Types of measurement errors

The way we deal with errors depends on their nature: sometimes errors cancel, and sometimes they require calibration and correction. There are two main types of error in experimental practice.

Additive errors $y = x + \varepsilon$

The error ε is independent of the magnitude of the true value x. It can be *systematic* when it causes a consistent shift in measured values away from the true value, defining measurement bias (miscalibration or environmental offsets); or it can be *random* and fluctuate unpredictably, defining the variance. Typically, random errors have zero mean, and the variance remains constant across the measurement range, so the effect of random noise is reduced by averaging repeated measurements (see signal stacking).[4]

Multiplicative errors $y = x(1 + \eta)$

The error η disturbs the true value x proportionally to the value itself. This type of error arises, for example, when the gain is not well calibrated or drifts with temperature. Note that taking logarithms converts multiplicative noise into additive noise: $\log y = \log x + \log(1+\eta) \approx \log x + \eta$ where the last approximation applies for small errors $\eta \ll 1$.

Additive and multiplicative errors propagate differently through models, bias estimators in different ways and call for different statistical treatments.

Statistical descriptors: center, spread and structure

Three salient characteristics of a given dataset determine its reliability: the mean or central tendency, the scale of dispersion and the structure of the dispersion, particularly when extreme values are critical.

Central tendency

Mean μ is the arithmetic average obtained by summing all values x_i and dividing by their number n. It is sensitive to every observation, including outliers.

$$\mu = \frac{1}{n}\sum_{i=1}^{n} x_i$$

Log-mean $\mu_{\log}$ may be more meaningful for positive data spanning orders of magnitude.

$$\mu_{\log} = \frac{1}{n}\sum_{i=1}^{n} \log x_i$$

Geometric mean (GM) is a multiplicative average and equals $\mu_{\log} = \log(\mathrm{GM})$

$$\mathrm{GM} = \left(\prod_{i=1}^{n} x_i\right)^{1/n}$$

Median is the value that lies in the middle of a ranked dataset, with half of the data above and half below. It is less affected by extreme values and provides a robust sense of the typical case.

Mode is the most frequent value in a dataset, useful for categorical or discrete data where "most common" is more informative than numerical averaging.

Measures of dispersion

Central tendency alone can be misleading: two datasets may share the same mean but differ greatly in their precision. Measures of variability capture data dispersion:

Variance σ^2 is the average squared deviation of data points from the mean. It penalizes larger deviations more heavily.

$$\sigma^2 = \frac{1}{n}\sum_{i=1}^{n}(x_i - \mu)^2$$

Standard deviation σ is the square root of variance, expressed in the same units as the data, making it easier to interpret in practice.

$$\sigma = \sqrt{\sigma^2}$$

Coefficient of variation (CV) is the ratio between the standard deviation and the mean. This dimensionless value allows comparisons across datasets with different scales. Widely used in quality control, it reflects relative consistency: the smaller the CV, the more reliable the process.

$$\mathrm{CV} = \frac{\sigma}{\mu}$$

Distributions

A probability density function (PDF) indicates how likely different values are. Distributional assumptions can be tricky. The Gaussian distribution is the workhorse of statistics, yet many datasets are inherently positive-valued (physical quantities), skewed, truncated or heavy-tailed. Selecting an appropriate probability distribution is therefore important for sound data interpretation and decision-making. Consider:

Uniform: all values are bounded within an interval and are equally likely, such as a random phase (between 0° and 360°) or position in a box.

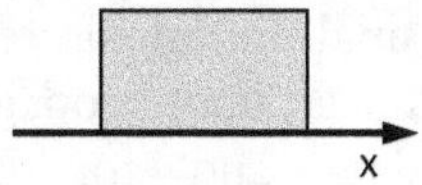

Normal (Gaussian): values disperse symmetrically around the mean and extend from –∞ to +∞.

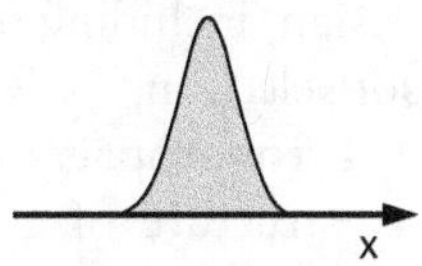

Log-normal: common for positive-valued physical parameters such as mass, age, permeability or material properties when the logarithm of the parameter follows a Gaussian distribution.

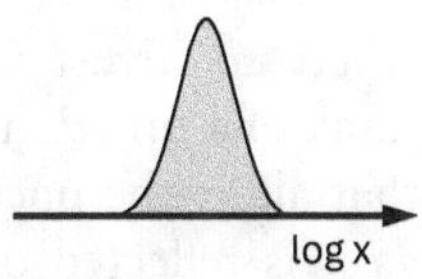

Poisson: describes discrete, rare events (e.g., radioactive decay, number of cracks in a specimen and earthquakes).

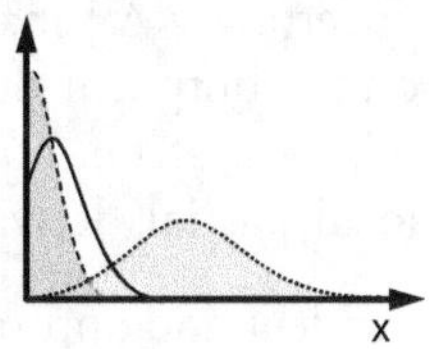

Heavy-tailed: In some systems, extreme values occur more often than the normal distribution predicts and may govern decision-making. In such cases, consider alternative distributions such as log-normal, power laws, Zipf's law or Lévy-stable distributions.

The Central Limit Theorem in experimentation

This is one of the most powerful results in statistics. The theorem states that if we take random samples of size n from a population with mean μ and standard deviation σ, then for sufficiently large n: (1) the distribution of the sample means $\bar{x}$ is approximately normal, regardless of the distribution of the original population; (2) the expected mean value of the sample means equals the population mean, $E\left[\bar{x}_i\right] = \mu$; and (3) the standard deviation of the sample means is $\mathrm{sd}\left(\bar{x}_i\right) = \sigma / \sqrt{n}$.

These results justify the widespread use of averages in noisy experiments and guide experimental design, such as choosing sample size or the number of repetitions needed to reduce noise through averaging.[4] However, important assumptions underlie the theorem: independence, finite variance and adequate sample size. When distributions are heavy-tailed, strongly skewed or correlated, the convergence may be slow or even misleading.

Uncertainty propagation

Errors interact with the system under study, sometimes canceling harmlessly and sometimes amplifying catastrophically, as in nonlinear systems where small initial discrepancies can dominate the response. Understanding how uncertainty propagates is therefore critically important both for interpreting data and for guiding experimental design, including noise control[5] and sensor selection.[6]

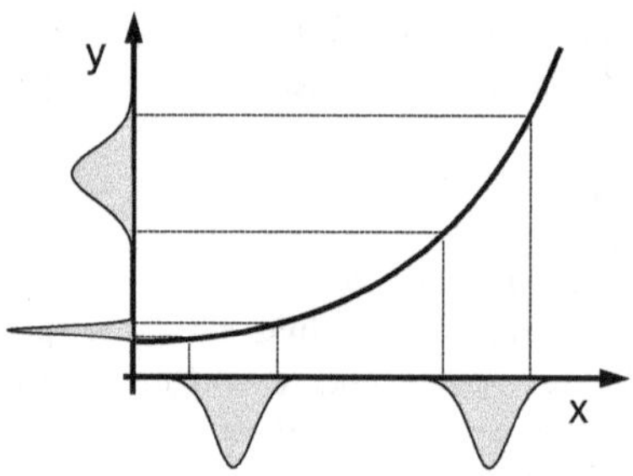

Error propagation reveals the sensitivity structure of a system, shows which variables matter most and indicates where to invest in precision. One can think of a function as a distorting mirror that alters the uncertainty in an input x as it is reflected onto the output y. An uncertainty Δx may have a minor effect on the output if the mirror is "flat" (low slope), but will be amplified if the mirror is steep. For multivariate functions, this steepness is captured by the local partial derivatives $\partial y / \partial x_i$. When input uncertainties are small, independent and approximately Gaussian, their contributions to the output variance combine by Pythagorean addition: $\sigma_y^2 = \sum_i \left(\frac{\partial y}{\partial x_i} \right)^2 \sigma_{x_i}^2$.

For example, consider a pendulum with period $T = 2\pi\sqrt{\ell / g}$. The temperature-dependent cord length ℓ and the local gravity affect the accuracy of a "grandfather" pendulum clock. The error in the length Δ_ℓ propagates to an error in the period Δ_T according to the partial derivative $\partial T / \partial \ell = \pi / \sqrt{\ell g}$. Then, the proportional error in the period $\varepsilon_T = \Delta_T / T$ is related to the proportional error in length $\varepsilon_\ell = \Delta_\ell / \ell$ as $\varepsilon_T = \varepsilon_\ell / 2$. Similarly, a proportional error in local gravity $\varepsilon_g = \Delta_g / g$ causes a proportional error in the period $\varepsilon_T = -\varepsilon_g / 2$. If both uncertainties are small and independent, they combine in quadrature $\varepsilon_T = \frac{1}{2}\sqrt{\varepsilon_\ell^2 + \varepsilon_g^2}$.

This approach breaks down in complex systems or when distributions deviate strongly from Gaussian behavior. In such cases, *Monte Carlo simulations* become indispensable. Monte Carlo methods rely on repeated random sampling: we draw values of the input variables from their probability distributions, compute the output each time and repeat, gradually building a picture of the output's distribution. These simulations expose the full distribution of outcomes and reveal interactions and tail risks that analytic methods may suppress.

Thoughtfulness in the face of uncertainty

We often underestimate the uncertainty in our own measurements. High precision can create the illusion of accuracy, even when hidden biases go unrecognized. A disciplined approach to uncertainty requires thoughtful experimental design, adequate and representative sampling, explicit treatment of bias, and an understanding of both instrument limitations and system sensitivity.

Statistics can illuminate when used thoughtfully. It cannot transform poor data into reliable conclusions, nor can it compensate for flawed models. But when embedded within careful experimental practice and guided by physical insight, statistical analysis allows us to discern structure within a noisy reality and build trustworthy knowledge from imperfect measurements.

Exercises

1. Central Limit Theorem: Assume a population uniformly distributed between 0 and 10 $(\mu = 5, \sigma \approx 2.89)$. Generate 100 samples of size $n = 36$, compute the sample mean $\bar{x}$ for each, plot a histogram of the $\bar{x}$ values and compute the mean and standard deviation of the sample means. Repeat for $n = 100$ and $n = 400$. Do the results confirm the three properties of the Central Limit Theorem: (1) the distribution of $\bar{x}$ approaches normality, (2) the mean of $\bar{x}$ equals μ, and (3) $\sigma_{\bar{x}} = \sigma / \sqrt{n}$?
2. Resonance-based mass measurement: A single-degree-of-freedom system has resonant frequency $\omega_r = \sqrt{k / m}$, so $m = k / \omega_r^2$. Assume $CV_k = 10^{-4}$ and $CV_\omega = 10^{-8}$. Use (1) the analytical solution and (2) Monte Carlo simulations to estimate the CV of the measured mass CV_m.

Notes

1. A. Lang (1844–1912), Scottish writer and anthropologist.
2. See essay on *Statistical Concepts for Experimental Design and Execution (Section 3).*
3. See essay on *Biases in Decision-Making (Section 2).*
4. See essay on *Signal Processing and System Analysis (Section 5).*
5. See essay on *Noise... What a Pain! (Section 4).*
6. See essay on *Sensing the Experiment: Scientific Sensors (Section 4).*

Essay 44

DATA AND MODELS

Errors and L-norms

"Data! data! data!" he cried impatiently. "I can't make bricks without clay."[1]

Models are fitted to scattered data to extract meaningful insights from uncertain experimental results. The recovered model parameters depend critically on two decisions that are often underestimated: (1) how individual measurement errors are defined, and (2) how these errors are aggregated into a single metric that reflects the overall quality of the fit.[2]

Individual error: definitions

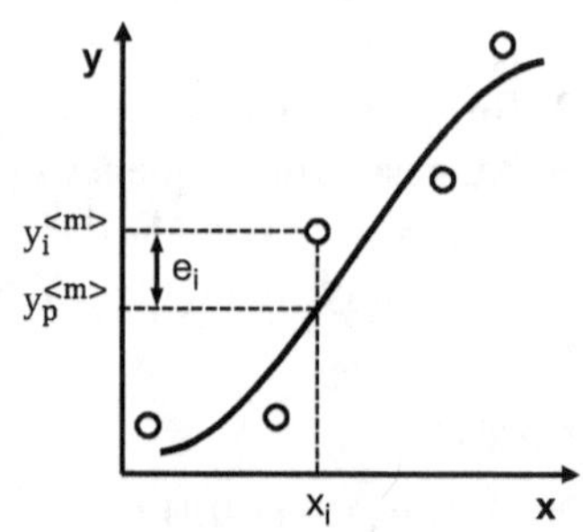

Given a dataset and a candidate model, the error or residual for the i-th measurement e_i is typically defined as the difference between the measured value $y_i^{\langle m \rangle}$ and the value $y_i^{\langle p \rangle}$ predicted from the model: $e_i = y_i^{\langle m \rangle} - y_i^{\langle p \rangle}$.[3] This linear error definition works well in many cases, particularly with additive noise ε. Challenges arise when errors are multiplicative η or when the data span multiple orders of magnitude. For example, when correlating personal income with energy consumption, small residuals at low incomes coexist with large residuals at high incomes, potentially causing the fit to be dominated by the wealthiest individuals. To balance contributions across the full range, alternative error definitions may be preferable:

DOI: 10.4324/9781003727521-49

Linear	$e_i = y_i^{\langle m \rangle} - y_i^{\langle p \rangle}$	Best for additive noise $y_i^{\langle m \rangle} = y_i^{\langle p \rangle} + \varepsilon$ Signed: the sign of the error captures over- and under-estimation
Logarithmic	$e_i = \log\left(y_i^{\langle m \rangle}\right) - \log\left(y_i^{\langle p \rangle}\right)$	Best for multiplicative noise $y_i^{\langle m \rangle} = y_i^{\langle p \rangle}(1+\eta)$
Proportional	$e_i = \dfrac{y_i^{\langle m \rangle} - y_i^{\langle p \rangle}}{y_i^{\langle p \rangle}}$	Also effective when y spans multiple orders of magnitude
Standardized (z-score)	$e_i = \dfrac{y_i^{\langle m \rangle} - y_i^{\langle p \rangle}}{\sigma_i}$	Accounts for the measurement of standard deviation σ_i May apply to both additive and multiplicative errors

A few quick diagnostic rules help identify the type of noise present in a dataset: (1) roughly constant linear errors (absolute values) suggest additive noise, (2) roughly constant proportional errors suggest multiplicative noise and (3) large errors at low y-values may indicate measurement resolution limits.

Error aggregation: norms

Once individual errors are defined, they must be combined into an overall measure of fit, commonly using the L_n-norm. Its definition and notable cases follow (Note: higher-order norms place greater emphasis on larger errors):

L_n-norm	$L_n = \left(\sum_i \lvert e_i \rvert^n\right)^{\frac{1}{n}}$	General definition Larger n values give more weight to larger errors
Sum of absolute errors	$L_1 = \sum_i \lvert e_i \rvert$	Democratic: all errors contribute equally Robust to outliers: good for noisy data
Sum of squared errors	$L_2 = \left(\sum_i \lvert e_i \rvert^2\right)^{\frac{1}{2}}$	Higher emphasis on larger errors Good when large errors are costly
Maximum absolute error	$L_\infty = \max \lvert e_i \rvert$	Focuses on the worst-case error Useful for safety-critical scenarios Highly sensitive to outliers

Fitting a model to uncertain data

The fitting process involves selecting a suitable model, ideally one grounded in physics rather than a generic polynomial, and adjusting its parameters to minimize the chosen error norm:

- $\min(L_1)$: the minimum total absolute error solution
- $\min(L_2)$: the least-squares solution
- $\min(L_\infty)$: the min-max solution (with a higher probability of non-unique solutions)

The least-squares solution performs best when errors are independent, have constant variance, and zero mean. Although Gaussian noise is often mentioned as a requirement, this mainly matters for statistical inference, such as hypothesis testing.

Consider a model with two parameters, a and b. The error norm defines a surface in the three-dimensional parameter-error space. A two-dimensional slice along one parameter, taken through the minimum error, reveals the convergence behavior and the extent of parameter resolvability.

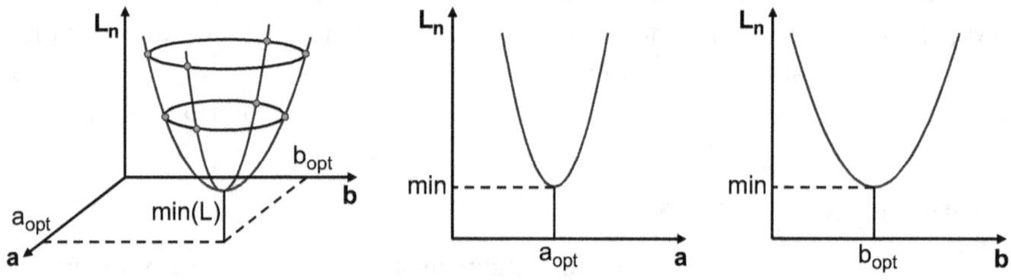

The combination of error definition and error norm determines the objective function. Different choices can yield different fitted parameters, even when the underlying model is the same. For example, least-squares fitting $\min(L_2)$ using linear errors versus logarithmic errors produces noticeably different trends and parameter estimates, illustrating the critical influence of these decisions (Figure: The dashed line shows the best fit based on logarithmic errors, whereas the solid line shows the best fit based on linear errors).

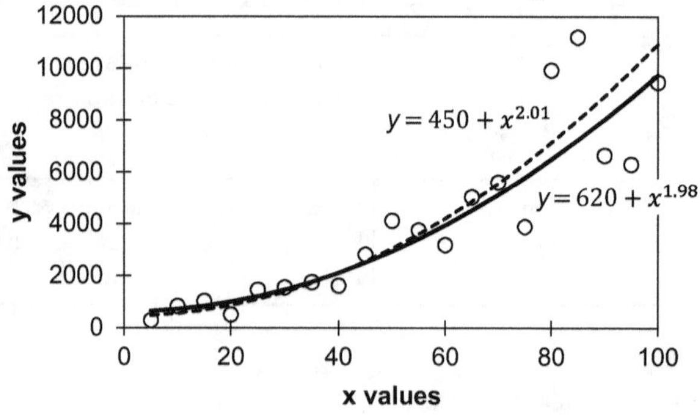

From data, through models, to new insights

Modeling data is the key to uncovering new insights from experimental measurements. Whenever possible, prefer physics-informed models over generic ones. Choose an error definition that matches the signal scale and the type of noise in your dataset, and select an error norm that balances the cost of large errors with robustness to outliers. Examine error surfaces to evaluate convergence and the resolvability of model parameters. Careful attention to these decisions ensures that the conclusions drawn from data are meaningful, reliable and truly representative of the underlying phenomenon.

Exercises

1 Create a synthetic dataset of 100 points based on a linear function $y = a + bx$ with added noise. Explore how the type of noise, error definition and error norm affect the fitted model parameters. Superimpose the various fitted models on the corresponding synthetic datasets and compare their sensitivity to outliers. Examine the error surfaces. Reassess the guidelines discussed in this essay.
2 Repeat the exercise using a real dataset from your field of interest.

Notes

1 S. Holmes in "The Adventure of the Copper Beeches" (1892) by A.C. Doyle (1859–1930), Scottish author.
2 Review essay on *Error Statistics and Propagation (Section 5).*
3 Note: errors can also be defined as the perpendicular distance to the trend line, which is advantageous for steep trends but is more complex to implement.

Essay 45

INFORMATION IN THE DATA

Science is built up of facts, as a house is built of stones; but an accumulation of facts is no more a science than a heap of stones is a house[1]

Advances in sensing, storage and communication have made data abundant and inexpensive. For example, a Formula 1 car in 2026 carries approximately 300 sensors and eight cameras, generating more than one terabyte of data over the course of a race. Yet in both racing and experimental research, value does not lie in the volume of data itself. What matters is information, the reduction in uncertainty and the deeper understanding that data make possible.

Lessons from the arts

The arts illustrate that complex systems can often be captured with far fewer parameters than their raw physical forms suggest. A musical score encodes an entire composition in a compact symbolic form; for example, F. Chopin's 90-sec *Minute Waltz* contains only ~2,000 notes, whereas a high-quality digital recording of the piece requires ~15 MB of data. Likewise, J. Vermeer's meticulous realism in *Girl with a Pearl Earring* and V. van Gogh's expressive brushwork in *Starry Night* convey vivid and meaningful scenes despite very different levels of visual detail.

These examples highlight a fundamental principle: information arises not from sheer data volume but from structured variation. Both excessive order (perfect regularity) and excessive disorder (pure noise) convey little information. Meaning emerges through the framework of prior knowledge, in a balanced regime between predictability and surprise, where patterns exist but

DOI: 10.4324/9781003727521-50

are not trivial. This principle applies as much to musical and pictorial expression[2] as it does to physical measurements.

Data compression

Digital compression methods make this relationship explicit and fundamentally limited by the signal statistics.[3] Compression algorithms exploit structure (redundancy, patterns and correlations revealed in transformed domains) and take advantage of human perceptual limits by discarding information that is inaudible or visually insignificant. JPEG, MP3 and modern codecs reduce storage by representing only the parameters that are relevant to the task, whether statistically or perceptually. The lesson for experimentalists is clear: not all measured data carry independent information.

Data-rich but information-poor?

Consider a toy tomographic problem in which an unknown region is discretized into four pixels with slownesses s_k. Suppose we record four travel times t_i from two horizontal rays (shown in the picture) and two vertical straight rays. These measurements lead to a linear system.

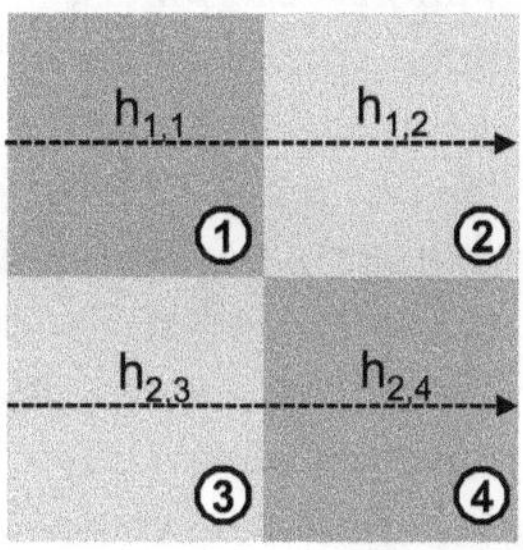

$$\begin{bmatrix} t_1 \\ t_2 \\ t_3 \\ t_4 \end{bmatrix} = \begin{bmatrix} h_{1,1} & h_{1,2} & 0 & 0 \\ 0 & 0 & h_{2,3} & h_{2,4} \\ h_{3,1} & 0 & h_{3,3} & 0 \\ 0 & h_{4,2} & 0 & h_{4,4} \end{bmatrix} \cdot \begin{bmatrix} s_1 \\ s_2 \\ s_3 \\ s_4 \end{bmatrix} \rightarrow \mathbf{t} = \mathbf{H}\mathbf{s}$$

where the transformation matrix **H** encodes the ray path lengths through each pixel. The four equations may appear distinct, yet the geometry forces one to be a linear combination of the others: the fourth row can be obtained by adding the first and second rows and subtracting the third. Consequently, the matrix **H** has rank three, not four (Reminder: rank is the number of linearly independent rows). This means that although four measurements were collected, they contain only three independent pieces of information.

In real experiments, rank alone is not the right metric. The appropriate analysis tool is singular value decomposition (SVD). SVD expresses the matrix in orthogonal modes whose strengths are given by the singular values, ordered from most to least constrained by the data: large singular values correspond to well-resolved components of the system, while small ones indicate directions in parameter space that amplify noise and are effectively irrecoverable.

Information-centered experimental design

Because the transformation matrix **H** encodes the sensor configuration (and the governing physics), the singular values can be examined *before* running an experiment. Thus, SVD becomes a valuable method for experimental design: it allows us to choose sensor placements, excitation patterns or measurement types that maximize the number of reliably recoverable parameters.

Information cannot increase as data pass through a noisy channel.[3] It can be preserved only if the transmission rate remains below the channel's capacity; if the rate exceeds this limit, information is inevitably lost. Noise and bandwidth are not afterthoughts in experimental studies: they fundamentally constrain what can be known.

Information vs. illusion

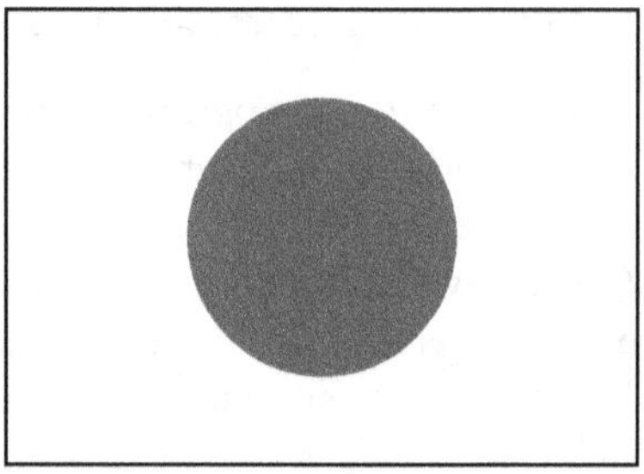

A raw digital photograph may contain millions of pixels, yet most natural images occupy a remarkably low-dimensional space. An extreme example is the flag of Japan: although a high-resolution photograph may require tens of megabytes, the underlying descriptive parameters are few – essentially the geometry and two colors.

Yet the converse is also true: incoherent or partially missing data can still yield deep structure when the underlying phenomenon is organized. The periodic table of elements emerged from incomplete and noisy measurements, but the coherence of atomic structure made the pattern discoverable.[4]

Modern machine learning algorithms trained on large datasets extend this capability further, revealing hidden structures that are inaccessible to manual inspection. In doing so, they have transformed the diagnosis of diabetes-related blindness, the detection of arrhythmias from electrocardiograms and the recognition of rare diseases.

Trained models can even infer plausible missing elements in data. A vivid example is image super-resolution, in which a deep learning model trained on a large dataset can predict high-frequency details that are absent in a low-resolution image.

Yet data inversion and machine learning carry a subtle danger. Models cannot generate reliable information where none exists, so when they attempt to infer beyond the information truly present, they can produce artifacts. In tomography, these appear as image ghosts. In machine learning, they cause hallucinations that are statistically consistent with the training data but do not reflect reality. In both cases, such artifacts can have significant implications for scientific inference and engineering decision-making.

In closing: data for understanding

Information is the reduction of uncertainty: the difference between what we know before an experiment and what we know after interpreting the data.[5] The modern proliferation of sensors and computational power tempts us to measure everything, but more data does not automatically yield more information. We must ask: *Which parameters are we trying to resolve? Are they recoverable? Does our experimental design make them observable?*

Data becomes information only when organized into patterns; information becomes knowledge only when those patterns are validated and integrated into our understanding. The task of the experimental researcher is therefore not only to accumulate data but also to design measurements, models and analyses that maximize recoverable information and deepen insight.

Exercises

1 Explore the Impressionists' style. Take a high-resolution image of a familiar scene with a well-known object (e.g., a bicycle). Reduce its resolution step by step (e.g., replace each 2×2 pixel block with its average). How does the information content change? At what point is the perceived "essence" lost? Discuss perception versus information.

2 Apply AI super-resolution to the images from Exercise 1. Compare the results with the original image. Which details are accurate, and which are "hallucinated"? How could similar effects impact scientific data analysis and inference?

3 Revisit the tomographic problem discussed above. Gradually add diagonal rays. Compute the transformation matrix **H** and perform SVD for each setup. What is the simplest experimental design that allows recovery of all four pixel values?

Notes

1 H. Poincaré (1854–1912), French mathematician and physicist.

2 R. Arnheim (1971). *Entropy and Art.*

3 C. Shannon (1916–2001), American mathematician; advanced the foundations of information theory.

4 See essay on *Knowledge Creation Throughout History (Section 1).*

5 A. Tarantola (2005). *Inverse Problem Theory and Methods for Model Parameter Estimation*, SIAM, 342 pages

Essay 46

INVERSION

Inferring causes from effects

Life can only be understood backwards, but it must be lived forwards[1]

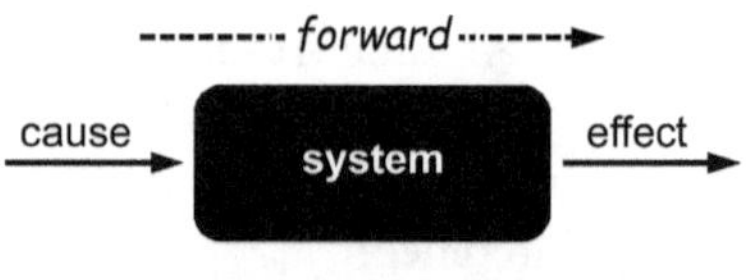

Forward problems begin with known inputs (causes) and predict outputs (effects) using a model of the system; in many systems, this is a convolution of the input with the system response.[2] In contrast, inverse problems start from observed effects and seek to determine either their causes or the properties of the system that produced them.

Forward and inverse problems are intertwined. J. Hadamard (1865–1963) noted that well-posed problems must satisfy three essential conditions: a solution must exist, it must be unique, and it must be stable, meaning that small perturbations in the input should lead to small changes in the output. While many forward problems satisfy these criteria, their inverse counterparts often violate one or more of them. Decades later, A. Tikhonov (1906–1993) introduced regularization methods to stabilize ill-posed inverse problems, laying the foundation for modern inverse solutions.

By the second half of the 20th century, inverse problems had become central to seismology and geophysics, enabling inference of the Earth's internal structure and guiding resource exploration. Medical imaging was transformed by X-ray tomography, culminating in the 1979 Nobel Prize awarded to A. Cormack for the underlying theory and to G. Hounsfield for the first practical computed tomography CT scanner. Today, inverse problems permeate

DOI: 10.4324/9781003727521-51

experimental research across engineering and science, from signal deconvolution and parameter estimation to multi-physics imaging in the laboratory, and from atmospheric science to anthropology and sociology in the field.[3]

Preliminary observations

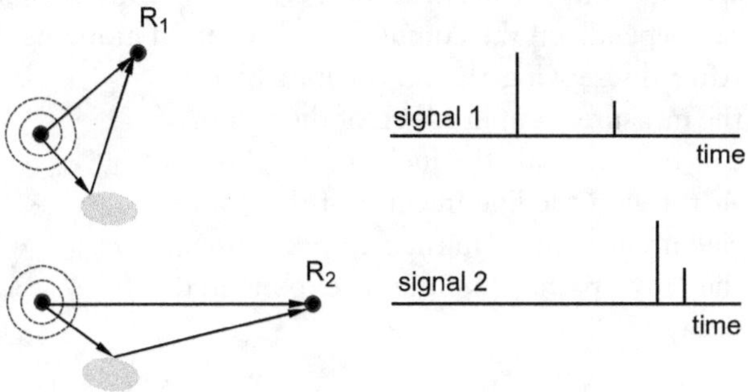

Consider an acoustic source that emits a sharp impulse. A receiver R_1 records a direct arrival followed by a second pulse, suggesting the presence of a reflector. However, the reflector's location is not uniquely determined with a single signal: many configurations could produce the same time signature. By relocating the receiver R_2 and repeating the experiment, additional information is obtained. Even when the location is inferred, measurement uncertainty, such as random timing errors, propagates into the computed reflector position.

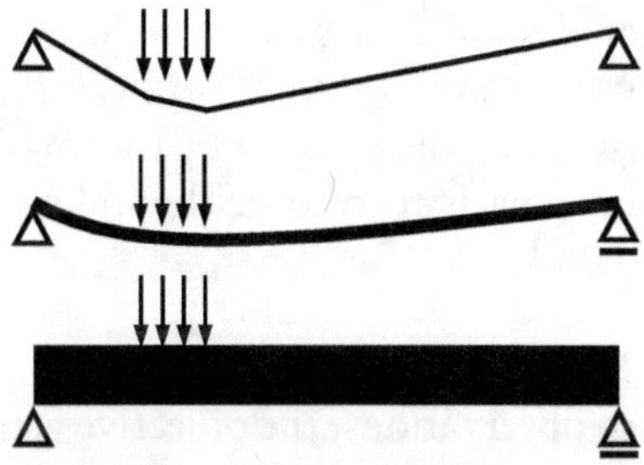

Some systems are more expressive than others, meaning they more clearly reflect the causes that generated the observed effects. For example, the deflection of a flexible cord (effect) is highly sensitive to the applied load distribution (cause), whereas a very stiff beam may exhibit deflections indistinguishable from measurement noise, preventing the inversion of measured deflections to estimate the applied load.

These examples illustrate the typical ill-posedness of inverse problems: the underlying system may obscure causal relationships, limited data can lead to non-unique solutions, and small measurement errors may produce large variations in the inferred parameters. These challenges correspond directly to Hadamard's criteria for well-posedness: existence, uniqueness and stability.

Inverse problems rely on a model to relate cause and effect, such as wave propagation in the acoustics problem, or elasticity theory in the beam example. Acceptable solutions must therefore be physically meaningful and explain the observations within the limits of the assumed model. In practice, inverse problems rarely yield a single definitive solution; instead, they provide a family of solutions that are consistent with the available data and any additional constraints.[4]

Inverse problems in matrix form

Many inverse problems can be written compactly in linear algebraic form. Two common examples are tomographic imaging and regression analysis.

Tomographic imaging uses measurements collected at the boundary to infer spatially varying material properties. In X-ray imaging, the attenuation of a ray depends on the cumulative absorption along its path through the object. After discretizing the object into pixels, the measured amplitude A_i of the i-th ray can be related to the incident amplitude A_0 through the line integral of the material attenuation μ_j in the j-th pixel along the ray path. Assuming exponential decay,[5]

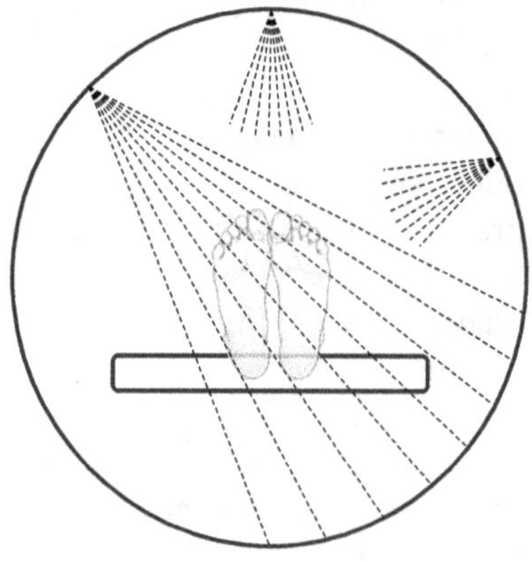

$$V_i = \ln\left(\frac{A_i}{A_0}\right) = \sum_{\text{source}}^{\text{receiver}} \ell_{i,j}\,\mu_j$$

where $\ell_{i,j}$ is the path length of the i-th ray within the j-th pixel. Writing such an equation for all rays and all projection angles yields a system of linear equations $\boldsymbol{V} = \boldsymbol{\ell\mu}$ in which the vector of m-measured values $\boldsymbol{V}\ (m \times 1)$ is related to the vector of n-pixel attenuations $\boldsymbol{\mu}$ $(n \times 1)$ by the matrix of path lengths $\boldsymbol{\ell}\ (m \times n)$. The objective of tomographic imaging is to recover the unknown attenuation field $\boldsymbol{\mu}$ from the measured data $\boldsymbol{V}$.

Regression analysis can also be viewed as an inverse problem. Given a set of observations, the objective is to infer the parameters of a model that best explain the data. Consider fitting a set of m elevation-time measurements zi, t_i using a polynomial function; for each measurement i, $t_i = c_0 z_i^0 + c_1 z_i^1 + c_2 z_i^2 + \cdots + c_{n-1} z_i^{n-1}$. Collecting all equations for the m-measurements yields a linear system of equations that can be expressed in matrix form as $\boldsymbol{t} = \boldsymbol{z}\,\boldsymbol{c}$ where $\boldsymbol{c}\ (n \times 1)$ contains the unknown model parameters, and both $\boldsymbol{t}\ (m \times 1)$ and $\boldsymbol{z}\ (m \times n)$ are constructed from the measured values.

Note that both tomography and regression express the measurements as a linear transformation of the unknown parameters $\mathbf{y} = \mathbf{h}\,\mathbf{x}$, where the vector $\mathbf{y}$ contains the measurements, the vector $\mathbf{x}$ contains the unknown parameters and the matrix **h** represents the transformation imposed by the experiment and the model. Direct inversion $\mathbf{x} = \mathbf{h}^{-1}\mathbf{y}$ is rarely feasible: the system may be overdetermined or underdetermined, the matrix **h** may be rank deficient, and even when an inverse exists, it can strongly amplify measurement noise.

Least squares solution

A common approach to solving linear inverse problems is the least squares method, which seeks the parameters **x** that minimizes the discrepancy between measured $\mathbf{y}^{\langle m \rangle}$ data and predicted values $\mathbf{y}^{\langle p \rangle} = \mathbf{h}\,\mathbf{x}$.[6] Then, the array of errors is $\mathbf{e} = \mathbf{y}^{\langle m \rangle} - \mathbf{h}\,\mathbf{x}$, and the sum of squared errors is $\mathbf{e}^T\mathbf{e}$. Let's expand it:

$$\mathbf{e}^T\mathbf{e} = \left(\mathbf{y}^{\langle m \rangle} - \mathbf{h}\,\mathbf{x}\right)^T \left(\mathbf{y}^{\langle m \rangle} - \mathbf{h}\,\mathbf{x}\right) = \mathbf{y}^{\langle m \rangle T}\mathbf{y}^{\langle m \rangle} - \mathbf{y}^{\langle m \rangle T}\mathbf{h}\,\mathbf{x} - \mathbf{x}^T\mathbf{h}^T\mathbf{y}^{\langle m \rangle} + \mathbf{x}^T\mathbf{h}^T\mathbf{h}\,\mathbf{x}$$

The least squares solution corresponds to the minimum point of the error surface. To find the minimum, let's take the derivative of the error surface with respect to **x** and set it equal to zero:

$$\frac{\partial\left(\mathbf{e}^T\mathbf{e}\right)}{\partial \underline{\mathbf{x}}} = 0 = 0 - \mathbf{h}^T\mathbf{y}^{\langle m \rangle} - \mathbf{h}^T\mathbf{y}^{\langle m \rangle} + 2\,\mathbf{h}^T\mathbf{h}\,\mathbf{x}$$

If $\mathbf{h}^T\mathbf{h}$ is invertible, the sought estimate of **x** is:

$\mathbf{x}^{\langle est \rangle} = \left(\mathbf{h}^T\mathbf{h}\right)^{-1}\mathbf{h}^T\mathbf{y}^{\langle m \rangle}$	*Least squares solution*

The matrix $\mathbf{h}^T\mathbf{h}$ is square $(n \times n)$ and symmetric, but it may not be invertible. A practical approach is to add a positive scalar η to the diagonal elements,

$\mathbf{x}^{\langle est \rangle} = \left(\mathbf{h}^T\mathbf{h} + \eta\,\mathbf{I}\right)^{-1}\mathbf{h}^T\mathbf{y}^{\langle m \rangle}$	*Damped least squares solution*

While this stabilizes the inversion and ensures a solution exists, the resulting estimate may be physically less interpretable, and the added damping η biases the solution away from the true underlying parameters. Better alternatives are described next.

Information!

At its core, inverse problem solving is about information: the information sought vs. the information contained in the data (and any additional constraints).[5] No inversion can extract more information than the data provide, and forcing a solution beyond this limit can lead to misleading or physically meaningless results. When a system is ill-conditioned, several strategies can be employed to stabilize the solution without over-interpreting the data:

Enhance the data

Increase the number of independent observations, improve measurement quality and reduce noise.

Reduce the number of unknowns – Favor simplicity (Ockham's Razor).[7]

We aim to extract as much information as possible from data, but indiscriminate detail is not always desirable. A robust approach in the examples above is to use larger pixels in tomography or a lower-degree polynomial in regression, avoiding overfitting. Problem parametrization also matters, as essential features can often be captured with far fewer parameters; recall the flag of Japan can be defined with only six descriptors, not millions of pixels.[5]

Add prior information about the solution – Tikhonov regularization

This is one of the most versatile methods for stabilizing ill-posed problems. For example, to obtain a smooth tomographic image, we can seek a balance between minimizing the data misfit $\mathbf{e}^{\mathrm{T}}\mathbf{e}$, and minimizing the roughness of the solution, measured by the second derivative of the image. The second-derivative operator is represented by the "regularization matrix" **R**. The resulting regularized least squares solution is

$$\mathbf{x}^{\langle \mathrm{est} \rangle} = \left(\mathbf{h}^{\mathrm{T}}\mathbf{h} + \lambda \mathbf{R}^{\mathrm{T}}\mathbf{R}\right)^{-1} \mathbf{h}^{\mathrm{T}}\mathbf{y}^{\langle \mathrm{m} \rangle} \qquad \textit{Regularized least squares solution}$$

where the regularization parameter λ is chosen to balance smoothness against the data misfit.

Other alternatives

The three equations framed above are commonly encountered in the inverse-problems literature, often with modifications to incorporate additional criteria, such as a prior estimate $\mathbf{x}^{\langle \mathrm{prior} \rangle}$, or to weight measurements according to their reliability. Other strategies include Bayesian approaches to incorporate prior knowledge of parameter distributions, constraints based on physical or empirical considerations, or truncated inversion to retain only components associated with the largest singular values.

Experimental design for inverse problems

Because the transformation matrix **h** depends solely on experimental design and data collection strategy, inverse-problem analysis enables optimization of experiments before data collection. Examining **h** reveals the number of independent measurements, poorly resolved parameters, potential noise amplification, and the relative merits of alternative designs. Therefore, conceive different experimental configurations and, for each configuration:

- Identify and tabulate the number of measurements m and the number of unknowns n.
- Analyze the matrix $\mathbf{h}^T\mathbf{h}$. Small diagonal elements invert into large factors, magnifying measurement errors.
- Compute the singular values of the transformation **h** and sort them from largest to smallest. Identify the singular value that is two to three orders of magnitude smaller than the largest; its rank indicates the amount of usable information contained in the transformation.[8]
- For configurations with a similar number of measurements m, favor those that yield evenly distributed data in the spatial and temporal experimental domains, and the largest number of significant singular values.
- Use insights gained from initial configurations to generate improved test designs, as needed.

Finally, implement the selected test configuration to acquire high-quality data by carefully selecting transducers and peripheral electronics, controlling noise and applying appropriate data preprocessing. Remove the influence of transducers and peripheral electronics from the measurements before data inversion.

What about neural networks used in machine learning?

Neural networks introduce a different approach to inverse problems. They typically involve a very large number of unknown parameters, including weights and biases, often far exceeding the number of training data points. In effect, neural networks trade strong prior assumptions and relatively few parameters (as in classical inverse problems) for weak assumptions and a large number of parameters. For a network with L layers and u neurons per layer, the number of unknowns is on the order of $n \approx L\,u^2$. For example, a relatively small neural network with two hidden layers and one output layer $(L = 3)$, each containing $u = 50$ neurons, has approximately $n \approx 7{,}500$ parameters.

Consequently, neural networks are usually highly underdetermined, with many more unknowns than training data points; a condition known as over-parameterization. Nevertheless, neural networks often perform remarkably well in practice; however, their effectiveness is not guaranteed. Successful applications often require large datasets, carefully designed architectures (e.g., smooth activation functions, appropriate depth vs. width), explicit or implicit regularization through training algorithms, and alignment between the model structure and the underlying physical process. Without these conditions, the claim that "neural networks work remarkably well" can quickly become wishful thinking.

Lessons learned

Traditionally, experiments have been guided primarily by physical intuition. Modern experimental design and instrumentation, however, should be conceived explicitly within the framework of inverse problem solving. This perspective applies equally to laboratory studies and field investigations.

As experiments progress, it is essential to consider: what information is truly needed, how expressive the experiment is with respect to the underlying causes, how much independent information the dataset contains, how sensitive the inferred parameters are to noise, whether the model captures the essential physics, and what additional information may be required to constrain the solution space. Well-designed and carefully implemented experiments maximize information content, minimize noise propagation, and yield stable inverse solutions at minimal cost.

There are many strategies for solving inverse problems, each balancing fidelity to the data against stability and physical plausibility. No method can recover information that the data do not provide.

This brief review of inverse problems highlights a broader lesson across science and engineering: complex phenomena can often be described by multiple plausible models, yet even persuasive explanations and good fits do not ensure understanding of the underlying causes. Recognizing the distinction between plausible explanation and physical reality is essential for rigorous research.

Exercises

1 Plot the dataset below and, by trial and error, find the coefficients of a third-degree polynomial that fits the data. Then compute the least squares solution using matrix inversion and compare the results.

x	0	0.1	0.2	0.3	0.4	0.5	0.6	0.7	0.8	0.9	1
y	1.14	0.86	0.56	0.08	0.08	0.04	−0.12	0.27	0.15	0.38	0.93

2 The raw signal from an NMR experiment shows the decaying magnetization M_i over time. It can be modeled as a sum of exponential decays: $M_i = \Sigma A_j\, e^{-t_i/T2_j}$. Formulate this problem in the standard linear system

form $\mathbf{y} = \mathbf{h}\,\mathbf{x}$ to extract the amplitudes A_j. Explore strategies for handling ill-conditioning and discuss implications.

3 Choose a problem in your field. Identify (a) measurements, (b) system/model and (c) unknowns. Express the problem mathematically and discretize as $\mathbf{y} = \mathbf{h}\,\mathbf{x}$. Discuss invertibility and analyze the implications for experimental design.

Notes

1 S. Kierkegaard (1813–1855), Danish theologist and existentialist philosopher.
2 See essay on *Signal Processing and System Analysis (Section 5).*
3 To expand on this essay, refer to Santamarina and Fratta (2005), *Signals and Inverse Problems*, J. Wiley (pdf available online).
4 Work by A. Tarantola (1949–2009), Spanish-French geophysicist.
5 See essay *Information in the Data (Section 5).*
6 See essay on *Data Models: Errors and L-Norms (Section 5).*
7 See essay on *Our Physical Universe: Laws and Principles (Section 1).*
8 This heuristic rule suggests that a reasonable condition number for a stable inverse solution lies between 10^2 and 10^3.

Essay 47

DIFFERENTIAL EQUATIONS

Processes in mathematical form (unstated assumptions)

The laws of nature are but the mathematical thoughts of God[1]

An experimentalist who avoids mathematical analysis is like someone who buys all the ingredients but never cooks a meal. Without equations, we are left merely describing what we observe, often aware that we may have witnessed something unprecedented yet unable to push the boundaries of understanding. It is a missed opportunity.

Consider processes in which parameters evolve over space or time. Changes in these parameters can be expressed as rates, such as a reaction rate $\Delta c / \Delta t$ or the rate of change of atmospheric pressure with elevation $\Delta p / \Delta z$. In such cases, analytical solutions naturally lead to differential equations. Let us explore this further.

Where do differential equations come from?

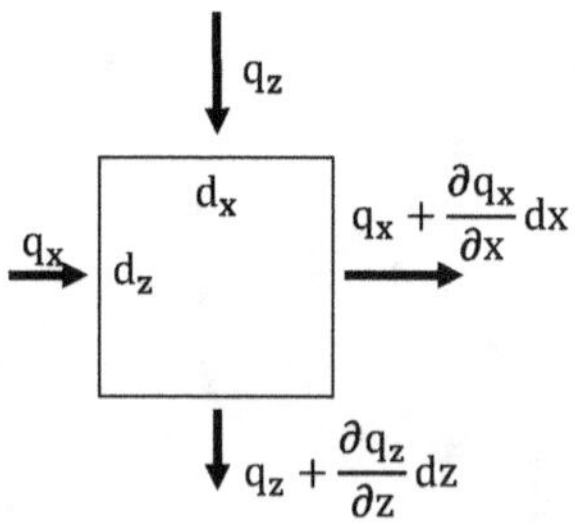

Most differential equations used in science and engineering (1) begin with conservation laws: mass, charge, momentum or energy; (2) assume that variables change smoothly; and (3) incorporate constitutive equations, often linear, such as Ohm's law $v = iR$, Newton's law $F = ma$ or Fick's law of diffusion $J = D\partial c / \partial x$.[2]

For example, consider planar flow through a porous medium in the x-z plane. The analysis focuses on a differential element of dimensions

DOI: 10.4324/9781003727521-52

dx [m] and dz [m]. The inflow q_{in} [m³/s] enters through two faces, while the outflow q_{out} exits through the opposite sides.

Assume *mass conservation*	$m_{in} = m_{out}$
Convert to flow rate by assuming the medium is saturated, perfectly rigid and the fluid is incompressible	$q_{in} = q_{out}$
Assume *variables change smoothly*. Estimate the outlet flow rate from the rate of change of flow in each direction: $q_x + \frac{\partial q_x}{\partial x} d_x$ and $q_z + \frac{\partial q_z}{\partial z} d_z$	$q_x d_z + q_z d_x = \left(q_x + \frac{\partial q_x}{\partial x} d_x \right) d_z + \left(q_z + \frac{\partial q_z}{\partial z} d_z \right) d_x$
Simplify to obtain	$\frac{\partial q_x}{\partial x} + \frac{\partial q_z}{\partial z} = 0$
Incorporate a *constitutive relation* that links response to the causal variable: Darcy's law, $q = ki$, relates flow rate to the spatial gradient in total energy, expressed here as head $i = \frac{\partial h}{\partial l}$	$\frac{\partial \left(k_x \frac{\partial h}{\partial x} \right)}{\partial x} + \frac{\partial \left(k_z \frac{\partial h}{\partial z} \right)}{\partial z} = 0$
Assume constant hydraulic conductivity k in both x and z	$k_x \frac{\partial^2 h}{\partial x^2} + k_z \frac{\partial^2 h}{\partial z^2} = 0$
Assume isotropy: $k_x = k_z$	$\frac{\partial^2 h}{\partial x^2} + \frac{\partial^2 h}{\partial z^2} = 0$

This final expression states that the divergence of the total energy is zero $\nabla h = 0$ when inflow equals outflow, and there is no net accumulation or loss within the system.

Common differential equations

We use the framework described above to revisit some common differential equations in engineering and science.

<u>*Newton's Second Law of Motion*</u>: $F = ma = m\frac{d^2x}{dt^2} = \frac{d(mv)}{dt}$

The relationship between a force F [N], the body mass m [kg] and the resulting acceleration a [m/s²] can be expressed as a second-order differential equation in terms of the second derivative of position x [m] with respect to time t [s], or equivalently, in terms of momentum conservation mv.

Wave equation: $\frac{\partial^2 u}{\partial t^2} = \frac{E}{\rho}\frac{\partial^2 u}{\partial x^2}$

A disturbance at one end of a rod propagates through the medium and is therefore not felt instantaneously at the other end. The wave equation arises from momentum conservation of a differential element (Newton's second law), elasticity as a constitutive law (Young's modulus E) and the relationship between strain and the smooth deformation field $u(x,t)$. This formulation applies to seismic waves, sound and ultrasound imaging. The electromagnetic wave equation has a similar structure, but its derivation begins with charge and energy conservation rather than mechanical momentum.

First-order kinetics equation: $r = \frac{dc}{dt} = -\alpha c$

In chemical reactions, mass is conserved but transferred from reactants to products. First-order kinetics assumes that the reaction rate r [mol/L·s] is proportional to the remaining reactant concentration c [mol/L], with α [s^{-1}] as the rate constant. Its solution $c = c_o e^{-\alpha t}$ describes the exponential decay characteristic of many natural processes.

Transport/advection equation: $\frac{\partial c}{\partial t} = v\frac{\partial c}{\partial x}$

Consider a scalar quantity $c(x,t)$ transported by a flow with velocity v [m/s]. Mass conservation requires that the instantaneous change in concentration at a differential element $\partial c / \partial t$ equals the difference in flux across the element $\partial(cv) / \partial x$. Assumptions include constant flow velocity, incompressible fluid, no sources or sinks, no diffusion or reaction and the scalar does not influence the flow. This equation applies to scalar quantities such as concentration or temperature.

Diffusion equation: $\frac{\partial T}{\partial t} = D_T \frac{\partial^2 T}{\partial x^2}$

For a metal rod heated at one end, thermal energy propagates toward the cooler end. The energy of a differential element depends on its mass density ρ [kg/m^3], specific heat c [J/kg·K] and temperature T [K]. Energy conservation implies that the rate of change of the element's thermal energy must balance the incoming and outgoing heat flux q_T [W/m^2]. The heat flux is determined by Fourier's law, $q_T = k_T \partial T / \partial x$. This yields the diffusion equation with thermal diffusivity $D_T = k_T / (\rho c)$ [m^2/s]. Assumptions include one-dimensional conduction, insulated lateral surfaces, homogeneous medium, uniform temperature across each cross-section, constant material properties and no internal heat sources or sinks (such as phase transformations). A similar derivation leads to the chemical diffusion equation, using mass conservation and Fick's first law, which linearly relates the concentration flux to the concentration gradient.

Alternative assumptions?

These examples show that deriving differential equations involves multiple simplifying assumptions. Relaxing these assumptions produces modified equations, for instance, by allowing spatial variability, anisotropy, reactive transport or nonlinear constitutive behavior. These new forms reveal additional physics and offer opportunities for improved modeling and data interpretation.

Differential equations make a difference

Differential equations are uniquely effective for describing processes. Their strength lies in their elegance, which yields insight from data, and in their robustness, which supports reliable predictions even beyond measured conditions. Each differential equation embodies distilled knowledge intertwined with assumptions; make those assumptions explicit to avoid misinterpreting results or drawing misleading conclusions. And above all, do not hold back... help the data express its full meaning.

Exercises

1 Choose the five differential equations most common in your field. For each equation, provide a detailed derivation, identify the starting point (the relevant conservation law), specify the constitutive relation used and list all assumptions, including implicit ones that may be easy to overlook.
2 Select a process relevant to your research and derive its governing differential equation, clearly stating the conservation law, the constitutive relation and all assumptions.
3 Derive the wave equation for particle motion $u(x,t)$ collinear with the propagation direction x. Assume a homogeneous, isotropic, linear-elastic, infinite continuum. Show the steps from force balance to the final PDE, state the constitutive relation employed (e.g., Hooke's law), list all assumptions and identify one explicit solution for $u(x,t)$.

Notes

1 Often attributed to Euclid (~300 BC), Greek mathematician known as the "Father of Geometry" for his work, *Elements*.
2 See essay *Models and Functions: Balancing Elegance and Reality (Section 5)*.

Essay 48

NUMERICAL METHODS

Power and peril

Machines take me by surprise with great frequency[1]

Experiments anchor us in reality. They reveal how systems behave under actual conditions and are indispensable for validating theories. Yet experiments are inevitably influenced by the messiness of the real world: imperfect boundaries, measurement noise and human error. They are often resource-intensive and, in many cases, conducting experiments may be impractical or impossible. In this context, numerical methods provide a complementary approach, but *they are not a substitute for reality.*

Motivation and advantages

Numerical methods allow the solution of systems of equations that rarely admit closed-form solutions. They are versatile and capable of handling complex boundary conditions, intricate constitutive models and coupled processes that would be intractable analytically. They enable exploration of parameter spaces, visualization of phenomena and upscaling of laboratory results to field conditions, provided assumptions and limitations are carefully considered.

Beyond exploratory analysis, numerical simulations can help optimize experimental design, support real-time test control,[2] facilitate data inversion and interpretation and extend experimental results into large datasets suitable for machine learning. They also enable the simulation of physically costly or otherwise impossible tests, including extreme time scales, length scales or environmental conditions. Most importantly, numerical methods provide a framework for testing hypotheses and generating new insights.

DOI: 10.4324/9781003727521-53

However, all of these benefits are contingent on careful attention to assumptions, discretization choices and input uncertainties.

Common characteristics and requirements

How do they work? Numerical methods replace a continuous problem with a finite, discrete and computable approximation. All methods involve three key steps. First, *discretization* converts governing laws into algebraic equations, derivatives into finite differences and integrals into weighted sums. Second, *approximation* is introduced, typically through linearizations or truncated series expansions. Third, *iterative computation* evolves the solution toward convergence.

Every algorithm must *consistently* approximate the correct problem, remain *stable* to prevent error growth, and *converge* toward the intended solution. A simulation may appear plausible, but without these safeguards, it can produce misleading results.

Discretized matter

We can explore matter at its elemental structure, such as the atomic scale for fluids and solids, or the particle scale for granular materials. In these methods, Newton's laws govern motion[3]: the displacement δ of an atom or particle with initial velocity v_0 over a given time interval Δt is $\delta = v_0 \Delta t + \frac{1}{2} a \Delta t^2$, where the element acceleration a results from the acting force F and its mass m, $a = F / m$. The algorithm evolves the discrete medium by integrating these equations of motion over time. The goal of these methods is not to simulate macroscopic phenomena or long-term processes but to gain insight into the underlying events at the elemental scale.

Discrete element method

DEM represents granular materials as discrete particles that interact through contact forces modeled with idealized contact laws, including Hertzian contact and Coulomb friction. Time integration proceeds in small increments, often on the order of microseconds or less. DEM can capture the complexity of particle-scale processes and emergent behaviors arising from local interactions, such as the evolution of force chains, dilation, shear localization and sequential particle breakage. However, the method is computationally intensive, limited to relatively small systems or short simulation durations, and sensitive to choices such as contact model, particle shape and time step. DEM is widely applied in powder technology, including pharmaceuticals and granular flow in silos, and can also be extended to other systems, such as crowd dynamics.

Molecular Dynamics

MD simulates the behavior of matter by evolving the motion of individual atoms through time. Atomic-scale attractive and repulsive interactions are computed from potentials such as Lennard-Jones. Time steps are typically femtoseconds, and simulations yield atomic configurations, estimates of material properties and insight into phenomena such as diffusion or protein folding. Yet MD is severely restricted in time and length scales, making macroscopic predictions or long-term processes practically infeasible without multi-scale coupling.

Discretized connectivity

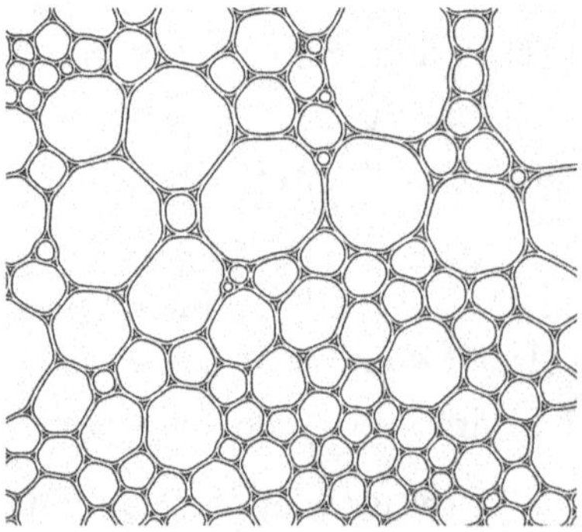

Some systems are governed by connectivity rather than local bulk behavior, including flow in foams or porous media, electrical grids, social networks and transportation systems.

Network models

NMs map complex macroscale systems onto an interconnected network, embedding governing laws and conservation principles within links and nodes. For example, studies of electrical conductivity through grids can use Ohm's law for links and charge conservation at nodes (Kirchhoff's laws). Similarly, liquid flow through a porous medium can be modeled using the Hagen-Poiseuille equation for tubular links and mass conservation at nodes.

Network models are intuitive and versatile, and they can leverage powerful tools from graph theory. However, the macroscale system response emerges from local interactions, and the accuracy of NM results depends strongly on the assumed network structure, the embedded physical laws, dynamic feedback such as those in reactive transport, heterogeneous interactions and stochastic processes.

Discretized fields

Macroscale simulations rely on equivalent continuum properties and field representations. Two common methods are described next.

Finite difference method

FDM approximates derivatives with finite differences. For example, the first and second spatial derivatives of a function $y(x)$ can be approximated as:

first derivative: $\frac{\partial y}{\partial x} \equiv \lim_{\Delta x \to 0} \frac{\Delta y}{\Delta x} \approx \frac{y_{i+1} - y_i}{\Delta x}$ | second derivative: $\frac{\partial^2 y}{\partial x^2} \approx \frac{y_{i+1} - 2y_i + y_{i-1}}{\Delta x^2}$

Similar expressions apply to functions of time $y(t)$ by replacing the spatial interval Δx with the time interval Δt. This approximation converts partial differential equations into algebraic equations. FDM is conceptually straightforward and computationally efficient for structured grids and simple geometries, but it struggles with steep gradients and stability constraints. Issues of consistency, stability and convergence are controlled by the order of the finite difference approximation, the spatial and temporal resolution of the grid, and the time or spatial stepping algorithm. FDM allows rapid exploration of physical systems and serves as the numerical backbone for many classical simulations in engineering and science.[4]

Finite element method

FEM solves differential equations by dividing a complex domain into discrete elements that form a mesh. The unknown field – such as displacement – is defined at nodes shared by neighboring elements. Within each element, the field is interpolated using shape functions.

For example, an elastic plate with a hole could be modeled as follows: (1) The plate is discretized with triangular elements, with a finer mesh near the hole where stress gradients are higher; (2) displacements are approximated by linear shape functions within each element, which results in constant strains inside the element; (3) strains define stresses and equilibrium is enforced across elements, while displacement continuity at nodes ensures compatibility; (4) the assembled algebraic system is solved for the nodal displacements under the prescribed boundary conditions; and (5) strains and stresses are then computed from the nodal displacements using the derivatives of the shape functions.

FEM excels for complex geometries, boundary conditions and nonlinear material behavior. Curved shapes, anisotropy, heterogeneity, nonlinear constitutive laws or large deformations can all be handled naturally within the formulation. Consequently, FEM underpins numerical macroscale analyses across disciplines, from mechanical systems and transport to electromagnetism. Choices of mesh design, element type and shape functions affect the solution accuracy and can yield convergence without correctness, which is a subtle but critical pitfall.

Final thoughts: power with caveats

The landscape of numerical methods is vast. This overview is intended as both an invitation to explore further and as a starting point for identifying tools suited to your specific research questions.

Numerical simulations are abstractions, not reality. They rely on assumptions about governing laws, material behavior and scale, and they inherit uncertainty from incomplete knowledge, imperfect inputs and boundary condition choices. These factors propagate through models and directly affect their fidelity and predictive power. Uncritical trust in simulation results, particularly for complex systems, can therefore be dangerous.

Validation is essential. Comparison with analytical solutions, benchmarking against simplified cases and isolation of individual processes help identify errors and limitations. While numerical modeling can clarify mechanisms, excessive parameter tuning to match experimental observations may mask compensating errors or overfitting, producing an illusion of accuracy. An agreement alone does not guarantee understanding.

For experimentalists, simulations are a powerful addition to the toolbox. Commercial and open source software can meet many practical needs, even without deep expertise in computational algorithms. However, conceptual understanding cannot be outsourced: a clear grasp of the underlying assumptions and limitations of numerical methods remains indispensable.

Numerical methods are powerful tools, but they are neither a substitute for experiments nor a shortcut. Their value depends entirely on the care, rigor and judgment you bring to their implementation. Used critically, they will sharpen your conclusions and expose phenomena that experiments alone cannot reveal. Used carelessly, they will mislead you with unwarranted confidence.

Exercises

1 What are the most common numerical methods used in your field of research? For each method, describe the techniques it employs for discretization, approximation and convergence.
2 Identify public-domain software that could be used to complement your experimental studies.

3 Human interactions can be loosely compared to particle or atomic interactions. Agent-based models (ABMs) adapt or extend DEM/MD-type algorithms to study crowd dynamics. Investigate the principles, assumptions and limitations of ABM in this context.

Notes

1 A.M. Turing (1912–1954), British mathematician and computer scientist.
2 See digital twins in the essay on *Devices: Design Strategies Enhanced by Digital Twins (Section 4).*
3 See essay on *Our Physical Universe: Laws and Principles (Section 1).*
4 See essay on *Differential Equations: Processes in Mathematical Form (Unstated Assumptions) (Section 5).*

SECTION 6

Technical communication

Essay 49

SCIENTIFIC COMMUNICATION

Discovery across minds

Great discoveries and improvements invariably involve the cooperation of many minds[1]

The research community is a living system, an intricate network of individual researchers, each with specialized roles yet constantly connected through publications, presentations and digital exchanges. Peer review and other forms of professional feedback serve as the community's immune system, preserving integrity and keeping the enterprise healthy. As conditions change, we evolve together, adapting, embracing breakthroughs and confronting new challenges.

Progress in science is not measured by the number of papers we publish or by citation metrics. For most of us, advancement comes through steady, cumulative contributions that mature over time, refine understanding and influence our field. In this way, scientific communication fosters both intellectual growth and societal impact.

Society sustains research so that it may, in turn, serve society. Therefore, communication must reach beyond the circle of specialists. Sharing our work through teaching, outreach and public dialogue allows ideas to travel beyond the laboratory and into the broader community. Charles Darwin's *On the Origin of Species* stands as a landmark example of research that transcended science and shaped culture itself.[2]

Scientific communication is intellectually demanding, inherently uncertain and profoundly transformative. Much of the literature focuses on its persuasive and rhetorical dimensions, showing how to reach, engage and move an audience. The essays in this section outline key principles and best practices

DOI: 10.4324/9781003727521-55

for crafting effective written and spoken technical communication, consistent with contemporary academic and cultural norms.

Communication is not merely the final step of a research effort: it is central to the research process itself. Writing and presenting force us to clarify our ideas, test the logic of our conclusions and expose our reasoning to the scrutiny of others. In doing so, we often discover new insights and inspire further inquiry, giving our findings a broader and more enduring life.

Do not read these essays as mere guidance for the final phase of your research. Begin writing and preparing presentations as soon as you begin your study. Communication is not just a means of dissemination. It enables discovery to happen, to matter and to last.

Exercises

1 Examine the communication style of one or more researchers you admire (via their papers, talks or recordings). For each researcher, identify and describe the key characteristics that define their style.
2 Identify science writers in popular media and analyze their work. For each writer, highlight the distinctive features of their communication style and note techniques that make their pieces accessible and engaging to a general audience.
3 Choose one of your own research findings and write two short descriptions: (a) for peers in your field – use technical terms and assume domain knowledge, and (b) for a non-specialist but educated audience – prioritize clarity and accessibility over technical detail. Then compare the two versions and write a brief reflection on how the emphasis, clarity and accessibility shift between them.

Notes

1 A.G. Bell (1847–1922), Scottish-American scientist and inventor of the telephone.
2 C. Darwin (1809–1882) hesitated to publish his work partly because evolution required an Earth far older than the accepted estimates of his time (In the 1650 *The Annals of the World*, Bishop J. Ussher had concluded that creation took place on October 23, 4004 BC; even by the 19th century, geological and physical estimates placed the Earth's age between only 20 million and 200 million years). Meanwhile, A.R. Wallace independently arrived at similar ideas about natural selection in his 1858 essay, prompting C. Darwin to finally publish *On the Origin of Species* in 1859.

Essay 50

PLAGIARISM

The subtle, the bad and the ugly

If it is not right, do not do it.
It is not true, do not say it[1]

We do not exist in a vacuum. In fact, we are inspired by previous work. In science and engineering, as well as in the arts and politics, idea generation often emerges through the combination and evolution of earlier concepts. There are numerous historical examples. G. Amontons' studies on friction influenced C.A. de Coulomb. E. Mach's ideas contributed to A. Einstein's theory of relativity. J. Bosch's surreal paintings inspired S. Dalí. T. North's translations and prose inspired W. Shakespeare's plays. And M. Gandhi's nonviolence shaped M.L. King's resistance.

The many faces of plagiarism

Our context and intellectual heritage are both necessary and undeniable. Yet there is a fine line between "gaining inspiration" and "copying". Let's explore a few recognizable and slippery paths to plagiarism:

- *The blatant plagiarist* copies sentences or paragraphs without alterations or references, often seen in early student reports and easily detectable using plagiarism detection software.
- *The pretender* extensively rephrases original content to disguise the source but preserves the underlying idea.
- *The misleading referencer* acknowledges others' work but in non-relevant parts of the text. This obscures the true influence of earlier work on the study.

DOI: 10.4324/9781003727521-56

- *The "make-happy" guilty-conscience writer* references a tangentially related paper by inspiring authors to appease them, while hiding their key work relevant to the current study.
- *The "defender of the oppressed"* highlights early, single-authored work – such as a thesis – but conceals more complete, peer-reviewed papers coauthored with mentors.
- *The oblivious co-author* unintentionally participates in plagiarism by being unaware that a collaborator omitted an essential reference.
- *The unethical concealer* avoids explicit plagiarism but deliberately leaves out closely related earlier work to exaggerate the novelty of the study.
- *The forgetful author* unconsciously reproduces an idea learned long ago and, having forgotten its origin, unintentionally presents it as new.
- *The self-loving author* excessively self-cites prior work to underplay or obscure relevant contributions from others.
- *The "long live the king" approach* cites a competing study only to highlight its weaknesses, withholds proper credit and treats it as a mere stepping stone to elevate the new work.

Do the right thing

Why do we do it? Sometimes it is sloppiness or laziness, sometimes ambition. Plagiarism ultimately reflects individual ethical lapses, but it is often exacerbated by institutional values and incentives: the rush to publish, pressure to perform, metric-driven evaluations, inadequate reward systems, additional compensation for publications and the rigorous demands for academic tenure all contribute but do not justify it.

What to do? Say NO to plagiarism. Above all, love the intellectual game. Focus on deep thinking and meaningful research to explore places where no one has been before. Do not ignore the little voice on your shoulder reminding you to acknowledge sources. Remember that proper citation does not diminish our own contributions but recognizes the shoulders we stand on.[2] Before submitting your work, double-check: *have you properly cited all sources you learned from?* For peace of mind, use plagiarism detection software, especially when collaborating or drawing from old notes. Just... *do the right thing!*[3]

And if you are on the receiving end? If you find that your work has been misrepresented, plagiarized or concealed, try not to dwell on it – it is not productive.[4] Frustration will only slow you down. Instead, channel your energy into what matters most: continuing to create and produce your best work!

Exercises

1 Write a brief essay for incoming students explaining what plagiarism is, why it matters and how different educational systems view it. Include clear examples and properly cite reliable sources.
2 Select several recently published academic articles and analyze their introductions using plagiarism detection software. Report similarity scores, discuss flagged matches and summarize what the results reveal about academic writing practices.

Notes

1 M. Aurelius (121–180 CE), Roman emperor and Stoic philosopher.
2 See essay on *A Great Question: Where Good Science Begins (Section 1).*
3 Film directed by S. Lee (1989).
4 Note: Some colleagues may feel morally compelled to denounce it.

Essay 51

SKETCHES AND PLOTS

Visual thinking and communication

A picture paints a thousand words...[1]

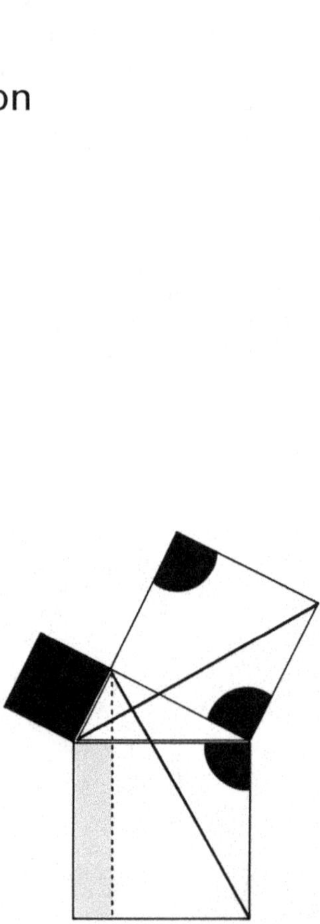

Euclid's *Elements* (c. 300 BCE) demonstrated the power of geometric diagrams as instruments of reasoning, including a rigorous proof of the Pythagorean theorem based entirely on visual logic.[2] Today, sketches and plots remain essential tools in research. They clarify concepts, reveal patterns, expose hidden processes and uncover causal relationships. At the same time, they serve as powerful means of communication, transforming abstract ideas into visual narratives that capture attention, convey insight and foster collective understanding. In both roles, graphical excellence lies in presenting complexity with clarity and integrity.[3,4]

Sketch: focusing attention and cognitive offloading

Sketches externalize evolving concepts onto a visual plane. By offloading ideas from the mind, sketching allows us to refine them iteratively while engaging the cognitive, motor and visual regions of the brain.

Elementary sketches readily convey the essence of an idea, but they must not be careless. Consider this sketch of a sharpened pencil[5]: can you spot details that contradict reality? Examine the conical shape of the lead tip, the

DOI: 10.4324/9781003727521-57

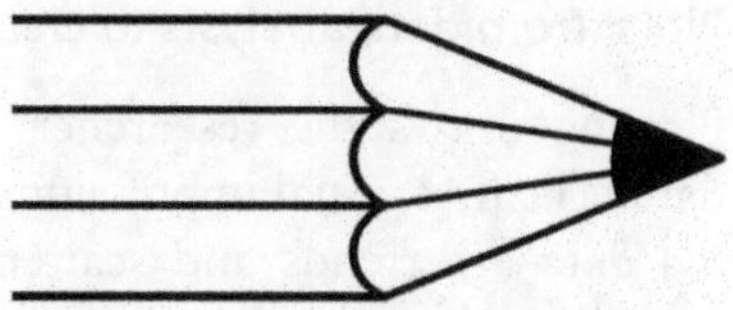

lines on the shaved wood, the direction of arcs along the ridges and the spacing between ridges. These discrepancies show how easily unchecked assumptions can intrude on our intuition and replace careful observation.

Sketching sharpens our attention. Try to draw from memory the car or bicycle you use every day. You will likely realize how little you actually know about its details. Observe it carefully, then draw it again.

As a research tool, sketching is dynamic. It begins with preliminary outlines of key concepts and evolves through successive refinements into physically accurate depictions that can ultimately guide mathematical formalism. For example, imagine the air–water interface gradually receding as water evaporates from a porous medium. An initial sketch may illustrate the general process, while a refined version incorporates capillarity, showing equal-radius menisci. Only a physically correct sketch can accurately capture air entry and reveal processes such as Haines jumps. L. da Vinci's progressive anatomical sketches, culminating in detailed drawings, provide a salient example of this process.[6]

Sketches have inherent constraints. Translating a three-dimensional reality into two dimensions requires conceptual abstraction. Thus, a sketch reflects deliberate choices about which aspects of a system to include or omit, shaping subsequent interpretations. For example, a pile of oranges may be drawn with the fruits in contact to study equilibrium or slightly separated to examine airflow.

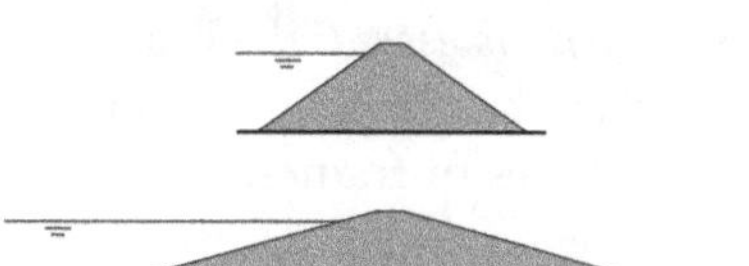

Finally, beware of physically incorrect sketches. They bias our perception of reality, distort analysis and hinder effective communication. Earth dams, often depicted with steep sides in textbooks, are much flatter to ensure stability. Sketches are not mere illustrations. They are cognitive aids and research tools that demand accuracy and thoughtful execution.

Plots: from data analysis to clear communication

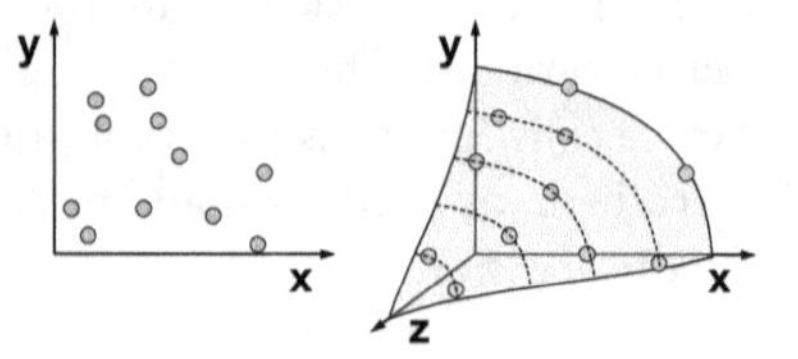

Plotting draws the researcher into the data. The first visual impressions of plotted data are trends and scatter. *Trends* stimulate the examination of correlations, the formulation of hypotheses and the exploration of potential causality. *Scatter*, or even a lack of correlation, prompts us not only to assess measurement errors but also to consider a missing variable, such as a z-variable in this sketch.[7]

During the early stages of visual data analysis, changing the scale of a plot is a valuable strategy for revealing underlying processes. Linear scales $y = ax + b$ are suitable when variables vary within one order of magnitude and relationships are linear. Use logarithmic or power scales when variables span several orders of magnitude, remembering that log(0) is undefined and the argument must be dimensionless. Logarithmic transformations linearize exponential relationships $y = ab^x$ in log-linear plots and power law relationships $y = ax^b$ in log-log plots, but they require careful interpretation to ensure accurate analysis. In general, avoid auto-scaling and maintain consistency across plots to facilitate the comparison of multiple datasets.

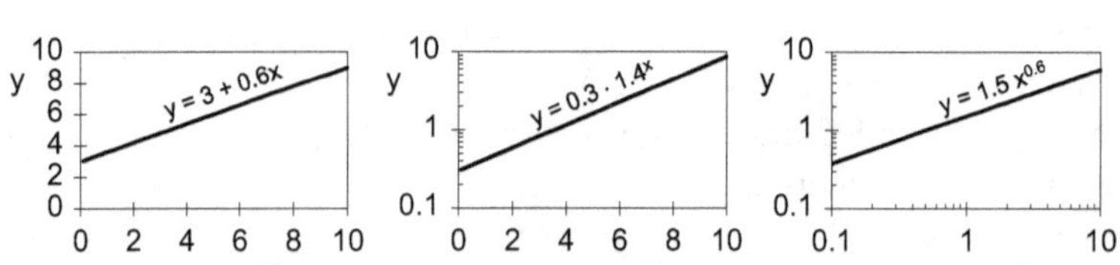

As a plot becomes increasingly refined through successive iterations, it gradually shifts focus toward effectively conveying insights to the target audience. The following recommendations can help create well-crafted, accurate and insightful plots[8]:

- *Clarity is essential.* Clean, uncluttered plots direct the reader's attention to what matters most. Avoid unnecessary components, including excessive gridlines or frames.
- *Legends.* Label datasets directly on the plot, and use figure captions to provide additional details. Detached legends force the reader to shift attention back and forth between the legend and the data, slowing comprehension and increasing the risk of misinterpretation.
- *Axes.* Clearly label each axis with the variable name, symbol and units, for example, "Wave speed, V [m/s]". Use actual values rather than percentages and consistent decimal precision.

- *Fonts*. Text should be large enough to remain readable even after resizing during journal proof preparation.
- *Data points or lines?* Use data points to represent measured values unless they are too numerous to distinguish, and use lines to represent models or theoretical predictions. Avoid automatically connecting points unless appropriate.
- *Color*. Color can add an extra dimension to a plot, but it should be used strategically. Make sure that data and trends remain distinguishable in black and white and maintain sufficient contrast to ensure accessibility.
- *Proportion*. Plot dimensions should be visually balanced. A golden rectangle with a 1.6:1 length-to-width ratio is generally pleasing to the eye, while a 1:1 square frame can emphasize similarity between variables. Directional data are best represented using a polar plot. When including multiple plots in a single figure, align them carefully and share axes whenever possible.

Closing thoughts

Traditional printed graphs provide a static snapshot. Today, these include artistic *scientific illustrations*, *graphical abstracts* that summarize key findings, and *infographics* that combine visuals with minimal text. Modern visualization also encompasses *interactive displays* that enable real-time data exploration and *immersive environments*.

From classical to contemporary techniques, sketches and plots remain essential tools for reasoning and communication in research, reflecting the central role of visual imagery in how our brains encode knowledge. Keep visualizations simple, clean and accurate. Spending a few extra minutes refining a plot can greatly enhance its clarity and impact – Always take the time to do it well!

Exercises

1 Expand your sketching and plotting toolbox for experimental research. Study classic examples from E.R. Tufte.[4] Analyze plots in leading newspapers and magazines to identify what makes them effective. Explore data animations such as time-evolving bubble plots to reveal trends.[3] Find applications of tripartite plots (e.g., three-axis logarithmic charts to analyze dynamic systems). Create Sankey diagrams to visualize resource transfers, such as energy flow charts from LLNL.

2 Imagine a pencil that advances the lead on demand. Sketch your concept and refine it until it appears functional. Then, disassemble a real mechanical pencil, observe every detail and produce a precise technical sketch with specifications clear enough for a machinist to build (Fun fact: the first patent was issued in 1822).

3 Compare different scales for presenting your data (linear-linear, linear-log, log-log, linear-power, etc.) and choose the one that best highlights the underlying process.
4 Identify journals in your field that require graphical abstracts. Analyze common features, styles and visual conventions; then design a graphical abstract for a current project or publication.

Notes

1 Usually attributed to the newspaper editor A. Brisbane (1864–1936).
2 Image adapted from O. Byrne (1847). *The First Six Books of the Elements of Euclid.*
3 H. Rosling (1948–2017) emphasized data visualization to make statistics alive and intuitive. He extensively used dynamic plots of international data to explore and deliver the most persuasive presentations explaining development trends. See his TED talks online.
4 E.R. Tufte (1983). *The Visual Display of Quantitative Information.* Graphics Press.
5 See H. Petroski (1992). *The Pencil: A History of Design and Circumstance.* Knopf Publishing Group refers to ENR 1891.
6 L. da Vinci (1452–1519) – Refer to Essay on *Knowledge Creation Throughout History (Section 1).*
7 See essay on *Correlation or Causation? (Section 5).*
8 Numerous examples available on the companion website illustrate and expand upon these principles. Although each journal has its own formatting and style requirements, your goal should always be to communicate your research results clearly and effectively to your intended audience.

Essay 52
WRITING
Crafting meaning one piece at a time

Easy writing makes hard reading[1]

Writing is inseparable from research itself. On an *individual level*, it helps us organize ideas, identify logical gaps and distill insights. Writing complements visual displays: while visuals engage our mental imagery, writing operates through semantic mental encoding and lets us work directly with concepts and meaning. At the *collective level*, writing connects researchers, enabling others to build on our work and forming the dynamic network we know as the research community. So, *put it before them briefly so they will read it, clearly so they will appreciate it, picturesquely so they will remember it, and, above all, accurately so that they will be guided by its light.*[2]

Overall structure: sequencing and flow

Technical manuscripts follow a characteristic structure in which the sections fit together naturally, creating a coherent framework rather than rigid pigeonholes. While the exact format varies across disciplines, common sections include:

- *Title.* The first element readers and search engines encounter; it largely determines whether your paper will be read and cited. Be specific, emphasize the main focus and relevance, and consider using two-part titles. Ask yourself: *What words will potential readers use in their search?*
- *Abstract.* A self-contained summary that captures the essence of the entire paper – from context and methods to results, conclusions and implications. It should stand alone as a coherent narrative.

DOI: 10.4324/9781003727521-58

- *Introduction*. Establishes the context, background and relevance of the study, leading to its motivation and objectives. In some journals (e.g., research letters), the literature review is integrated within this section.
- *Previous Studies*. Reviews prior work to build the knowledge base and lead the reader to the boundary of current understanding. It may include an extensive collection of published data from the field and related disciplines. A thorough review helps identify knowledge gaps and open questions, guiding the present study while avoiding reinventing the wheel.[3]
- *Methods*. Describes the research approach in enough detail to ensure reproducibility. Include dimensional analysis materials, devices (sensor selection, cell design, noise control, calibration) and test protocols. Use figures to clarify complex setups.
- *Results*. Presents findings clearly and objectively through a balanced combination of text and figures. The narrative should guide the reader through the data, describing observations without interpretation.
- *Analyses*. Processes and interprets data; it may include statistical methods, physical reasoning, analytical solutions, numerical simulations, data inversion, parameter extraction and uncertainty evaluation.
- *Discussion and Implications*. Interprets the results within the broader context of prior work. Compare experimental, analytical and numerical findings. Explain the underlying reasoning, acknowledge limitations and answer the "so what?" question by clarifying the significance and implications of your study.
- *Conclusions*. Summarizes the key findings and contributions, highlighting advances beyond the shoulders of those who came before.[3] This section should leave readers with a clear sense of closure and insight.

Developing the manuscript: assembling an evolving puzzle

Writing translates thoughts and observations into a formal system of words organized by grammatical rules.[4] In doing so, we develop the manuscript sentence by sentence, gradually assembling a self-defining puzzle until a clear picture emerges.

The following practical guidelines are presented in chronological order, as if you were preparing a paper. They summarize best practices we have learned over the years and are intended to help you get started. With time, you will develop and embrace your own approach.

- *Figures tell the story*. Start with visuals. A reader should be able to grasp the essence of the paper through its figures and captions; furthermore, they should support the conclusions. Select figures and tables carefully, refining them to publication quality.[5] Include detailed captions that guide readers to the key aspects of a plot; they should provide a concise title, context and data sources and any notes necessary for comprehension. Focus on facts rather than conclusions.

- *Identify the skeleton.* With the story in mind, define the overall structure of the manuscript. The outline above is the starting point. Section titles and subtitles should create a logically sequenced, clear structure.
- *Write a complete first draft.*
  - o Begin with short paragraphs, each focused on a single complete idea: even excellent ideas can lose impact when delivered all at once. Rearrange content until a coherent story emerges. Remember, you are assembling a puzzle.
  - o Make natural phenomena – not authors – the subject of your sentences. For example, write "Turbulent eddies increase drag (Jones, 2022)" rather than "Jones (2022) observed that turbulent eddies increase drag".
  - o Keep your audience in mind. Imagine potential experts reading your paper – write as if speaking to them. But explain with clarity and without jargon so others can follow as well.
  - o Standardized or formulaic expressions signal lazy writing. Keep your tone fresh, engaging and strong.
  - o *The order around switch until it good sounds.* Experiment with word order until sentences sound natural and the text flows smoothly. Avoid overusing parenthetical remarks.
  - o No statement is completely true. Do not insist that your ideas are "the best"; let evidence and examples persuade the reader.
- *References.* Be meticulous. Cite only papers you have read, and verify that the information you attribute truly matches the source. Always give proper credit; never hide it.[6] Choose references carefully and present them in the right context. Citing famous names without true relevance creates a false appeal to authority.
- *Equations and variables.* Introduce variables with their name, symbol and units in square brackets. When an equation involves several new variables, list the key ones before presenting it; for example: "The inertial force F [N] depends on mass m [kg] and acceleration a [m/s²], $F = ma = \left(2E / v^2\right)a$, where E [J] is the kinetic energy and v [m] the velocity".

Editing: concise and clear

At this stage, the figures, tables and the main body are ready and well-structured. The next phase is careful editing and rewriting to enhance clarity and flow. Indeed, *the secret to good writing is good editing.*[7] Early revisions should refine the structure and logical sequencing; later ones should polish sentences. Strive for clear, economical text: omit the needless, avoid repetitions and skip fancy words.[8] The following guidelines will help turn a first draft into a finished manuscript.

- *Sentence structure and consistency*. Prefer the standard "Subject + Verb + Predicate" form, and maintain a single tense throughout. For example, using the present tense tells the story as if it were unfolding in real time.
- *Avoid common pitfalls*. Limit the use of passive voice (except perhaps in methods/procedures), excessive gerunds and compounded verbs (carry out → conduct). Avoid extreme qualifiers, informal language, jargon, clichés and words that delay comprehension (e.g., "respectively"). Be mindful of overusing "the" (use plurals instead) and "of" (the analysis of data → data analysis).
- *Punctuation*. Use punctuation to enhance clarity and precision, but avoid overuse. Be consistent.
- *Language*. Technical English varies across disciplines, but clarity is universal. Strive for precision. Avoid using multiple interchangeable terms for the same concept, or opinionated inserts such as "fortunately" or "interestingly". Vary your vocabulary and avoid repeating the same word within a few lines.
- *Keep it brief*. Brevity is essential for clarity. Avoid unnecessary words in a sentence, unnecessary sentences in a paragraph and unnecessary text in the manuscript. One of the most effective editing strategies is simply cutting! Technical writing leaves no room for repetition: say it once and say it right. Beware: short writing takes long thinking.[9]
- *The ear as editor*. Step away from your text before revising; distance sharpens perspective. Then read it aloud: awkward phrasing and inconsistencies reveal themselves more readily to the ear than to the eye. A well-tuned ear makes a better writer.
- *Easy fixes*. Focus on word choice, minor grammatical errors, misplaced clauses or phrases and wordiness. Short sentences are acceptable if they convey ideas clearly. Watch for excessive use of "we", often inserted to avoid the passive voice.
- *Harder fixes*. Address irrelevant or disorganized background information, overly dense discussions or conclusions and too many ideas packed into one sentence or paragraph. Aim for a reader-friendly text, even when the material is dry.

AI-based editing tools are now widely available. They can accelerate editing, refine sentence structure, catch grammatical errors and suggest more fluent alternatives, helping reduce barriers for non-native speakers. However, over-reliance on AI can lead to shallow reasoning, factual errors ("hallucinations") and a loss of intellectual ownership. Researchers who depend entirely on AI risk producing text that is polished but lacks depth, whereas dismissing AI altogether unnecessarily discards a valuable source of support. The best outcomes arise when researchers first articulate their own ideas, retaining full control over the reasoning process, then use AI to improve clarity and flow, and finally verify that the revisions preserve the original meaning and scientific accuracy. This final verification ensures that authors fully understand the implications of every AI suggestion. AI can assist, but it cannot replace the critical thinking and disciplined attention that scientific writing requires.

Critical: the end... and the beginning

Finally, let's take a closer look at the two most important parts of the paper: the *conclusions*, which distill the essential contributions of your work, and the *abstract*, which summarizes your study into an appealing story that draws the reader in.

Conclusions = Extracting the Essence

> "WHAT IS ESSENTIAL IS INVISIBLE TO THE EYE."
>
> —LE PETIT PRINCE

Avoid conclusions that merely restate what is obvious or already well established in the literature. Revisit your data and text to uncover what others may have missed. *What is essential is <often> invisible to the eye*[10]; these are the gems. Reread the paper with a fresh mind, highlight statements and findings that push the boundaries of knowledge, and organize them logically. Synthesize them into short, punchy sentences. Do not introduce new information or repeat results; instead, focus on insights, identify implications and emphasize broader significance. Finally, read your evolving conclusions critically: are all statements substantiated by the work? Refine until only genuine contributions remain.

Abstract: The Front Window

Think of the abstract as the paper's front window: it displays the essential elements of your work and entices readers to engage with the full story. It also informs search engines and helps others discover your paper. Write it at the very end, when you are ready to submit. For a first draft, copy the final manuscript into a new file, retain the most important statements from each section and progressively condense until you reach the target length. Then refine for logic and flow. Caution: abstracts generated by AI tools are often deficient; it is up to you to produce a coherent narrative that accurately captures your study in a self-contained, appealing story. As you finalize the paper, ask yourself: *What have I missed?* And reconsider the title once more!

There are still a few thoughts left in this ink bottle...

Writing is an integral part of research: clear thinking produces clear writing, and, in turn, writing stimulates clear thinking and deeper understanding. We learn to write by reading good writing, noticing what works and becoming our own toughest editors. Above all, writing takes time – even proficient writers may spend a full morning crafting a single decent page! Mastery comes through dedication: block time to write every day, sit down, resist distractions and make your writing environment a pleasant space.

Writing with coauthors? Iterations can become endless. Submit your very best version in each cycle to minimize revisions. And if in doubt, run contributions through plagiarism-check software; it may feel uncomfortable, but being associated with a careless publication damages credibility.

Words matter. A paper that reads smoothly but leaves no memory has failed. Conversely, substance alone is not enough: style brings a paper to life. Just as food is nourishing without spices but becomes memorable with them, a technically correct paper can be bland if it lacks the spark that makes its contributions stand out.

Make full use of the wonderful writing resources available to you, but never relinquish your intellectual power. Writing is the gym of the mind where ideas stretch, strengthen and take shape. It rewards discipline with discovery. Writing is creation – Enjoy it!

Exercises

1 This essay was particularly cumbersome to assemble; for example, the first section, on the structure of papers, risks preempting later sections that follow the chronological process of writing a paper. Analyze the essay in detail. Can you propose an alternative structure that improves clarity and flow?

2 Select one of your manuscripts and run successive reviews using the guidelines provided above. Identify elements that could be revised to enhance clarity, conciseness, coherence and overall flow.

3 Choose a paper from your field and examine its content carefully. Then: (a) suggest three possible alternative titles – discuss the merits and drawbacks of each; (b) analyze the abstract: does it stand alone? Does it adequately summarize the paper? Is it engaging and enticing to the reader?

4 Write a 250-word critical discussion of a well-known reference in your field. Identify errors, omissions and limitations, as well as significant insights. Can you extract an essential observation that the authors may have overlooked? Focus on substantive points that shape the scientific value of the work.

Notes

1 Generally attributed to R.B. Sheridan (1751–1816), Irish playwright, poet and politician.
2 J. Pulitzer (1847–1911), Hungarian-American politician and publisher, known for the Pulitzer prize in journalism, arts and letters.
3 See essay on *Literature Review: Mapping the Knowledge Landscape (Section 1).*
4 N. Bohr (1885–1962), Danish physicist, was keenly aware of this challenge as he sought to express quantum physics in plain language.
5 See essay on *Sketches and Plots: Visual Thinking and Communication (Section 6).*
6 See essay on *Plagiarism: The Subtle, the Bad and the Ugly (Section 6).*
7 H. Guinness, Irish writer and photographer, quoted in a 2020 *New York Times* article titled "How to Edit Your Own Writing".
8 W. Strunk Jr and E.B. White (1959). *The Elements of Style.* New York: Macmillan – a concise, enduring guide to good English writing, famously used and praised by J.F. Kennedy.
9 Quoting B. Pascal (1623–1662), French mathematician and physicist: "I have made this letter longer than usual because I have not had the time to make it shorter".
10 A. Saint-Exupéry (1943). *Le Petit Prince.* New York: Reynal & Hitchcock.

Essay 53

PUBLISHING

Join the scientific conversation

Real isn't how you are made...
You become. It takes a long time[1]

To publish and be cited is, in a sense, to exist in the history of science and engineering: "I exist only insofar as I exist for others".[2] For some giants, like K. Gödel,[3] a small number of publications secured a lasting legacy. Yet bibliometric studies show that highly productive researchers are also more likely to produce high-impact work.[4] Indeed, many great minds were not only brilliant but prolific; L. Euler, for example, authored hundreds of pioneering contributions.[5]

Time and energy are limited – use them wisely. Publish research that advances understanding or offers practical solutions to meaningful problems. Choose themes that matter, ideally with implications across fields. Strive to be both innovative and thorough. Submit work that is complete and convincing, grounded in solid theory, sound data and a broad awareness of prior research. Write clearly and concisely. Quality papers endure and stand the test of time.[6]

The process of publishing

Journal selection. Choosing where to publish is an exercise in identifying your intended audience. A good match places your work within an active conversation and ensures it reaches readers who can build on it. Aim for the strongest journals that align with your readership, but remember that great papers can thrive regardless of a journal's impact factor. A practical hint: if you rarely cite the journal you are targeting, it may not be the right audience for your work.

DOI: 10.4324/9781003727521-59

Journal formatting. Ensure that your manuscript adheres to the journal's style: structure, length and references. Read the "Instructions for Authors" carefully and examine recently published articles to understand their tone, scope and expectations.

Submission. Submit only your best work. Sending a manuscript before it is ready is unprofessional: it wastes reviewers' time, sets a poor precedent and undermines your own effort.

Responding to reviewers. When reviews arrive, read them carefully to understand what is being asked. A useful strategy is to prepare a two-column reply: reviewer comments on the left, your responses on the right. You do not need to agree with every suggestion, but address each one transparently, accommodating what you can while preserving scientific rigor. Do not rush your response; be meticulous and stay truthful to the science.

The revised manuscript. Revisions can fragment a well-structured paper: you may satisfy reviewers yet lose coherence in the process. After making changes, step back and reread the manuscript as a whole. Check the flow, logical sequence and narrative clarity. The story should remain engaging while staying true to the original narrative.

After publication. Acceptance is only the beginning. Share your paper with colleagues, present it in talks and create supplementary materials such as short videos or training modules to extend its reach. Social media can amplify visibility, but avoid self-promotion – it is rarely well received. The goal is not to inflate metrics but to foster meaningful exchange.

Co-authorship

Collaboration is often synergistic and sometimes essential to advancing research. A strong team (small enough to stay focused yet diverse enough to broaden perspectives and skills) strengthens the work. Strong collaborators also bring critical eyes that refine ideas and improve quality.[7]

At the same time, co-authorship presents challenges: keeping track of drafts, reconciling writing styles, balancing workloads and managing personalities, delays and sensitivities. Each contributor tends to value their own effort more highly than others do, and feelings of ownership often shift as projects evolve. To avoid misunderstandings, discuss authorship early, before tensions arise. Be fair and ethical, and include only those who have made genuine intellectual contributions.

Grow as an author

The papers you write early in your career matter more than you may realize. They set the tone for how others perceive your work and form the foundation on which your future research will grow. Good publications foster a reputation for rigor, clarity and creativity. A strong reputation, in turn, opens doors to better collaborations and opportunities, richer environments, and ultimately, better science. Remember that good papers arise less from sudden flashes of brilliance than from steady habits. Writing regularly keeps ideas alive.[8]

Once published, papers take on lives of their own. Some shine briefly, drawing attention for a few years before fading. Others prove resilient, continuing to shape discussions decades later. A few are rediscovered in new contexts, their relevance unfolding again. Occasionally, a paper lies dormant until its time finally arrives. Citations may rise and reputations may grow, but the true measure of your work is whether it helps others think differently, discover new dimensions or solve problems that matter.

Publishing is not about scattering many small pieces that add little; it is about building a body of knowledge. A coherent sequence of contributions allows ideas to mature, methods to evolve and understanding to deepen. What matters is not how many papers carry your name, but whether your work truly moves the field forward. Eventually, you become *Real*:

> *Real isn't how you are made. It's a thing that happens to you. It doesn't happen all at once. You become. It takes a long time. That's why it doesn't happen often to people who break easily, or have sharp edges, or who have to be carefully kept. Generally, by the time you are Real, most of your hair has been loved off, and your eyes drop out and you get loose in the joints and very shabby. But these things don't matter at all, because once you are Real you can't be ugly, except to people who don't understand.*[1]

Exercises

Select the five most influential papers in your field from the past 30–50 years. For each paper,

- Identify the paper's key contributions to the field.
- Analyze the writing style and presentation.
- Use bibliometric tools to examine the paper's citation history and trends.
- Generate citation maps that visualize its connections to other works.
- Discuss how ideas from the paper have influenced the evolution of the field.

Notes

1 M. Williams Bianco (1881–1944), English-American author of children's books. Edited from *The Velveteen Rabbit.*
2 E. Mounier (1905–1955), French philosopher on relational personalism. The original quote is: "*Je n'existe que dans la mesure où j'existe pour autrui*".
3 K. Gödel (1906–1978). Born in what is now the Czech Republic, he was one of the most influential logicians of the 20th century. He is best known for his 1931 paper on formally undecidable propositions.
4 P. van den Besselaar and U. Sandström (2015). They found that in a sample of 48,000 researchers, the most productive individuals were significantly more likely to produce top-cited papers.
5 L. Euler (1707–1783), Swiss mathematician, physicist and engineer who authored 866 publications.
6 Review previous essays: *A Great Question: Where Good Science Begins (Section 1), Literature Review: Mapping the Knowledge Landscape (Section 1), Sketches and Plots: Visual Thinking and Communication (Section 6)* and *Writing: Crafting Meaning One Piece at a Time (Section 6).*
7 See essay on *Research Teams: Creative Synergy (Section 2).*
8 See essay on *Literature Review: Mapping the Knowledge Landscape (Section 1).*

Essay 54

PRESENTING

Tell the story - Inspire

It usually takes me more than three weeks
to prepare a good impromptu speech[1]

A well-prepared talk can captivate an audience, make complex ideas accessible and leave a lasting impression. Yet many presentations fall flat – not because the content is weak, but because the central message is unclear or the story is missing. As you prepare, ask yourself: *What do I want the audience to remember once the talk is over?* Every element (structure, visuals, timing and even body language) should work together to serve that single purpose.

Presentation styles vary across disciplines. While some exceptional talks succeed without slides, presentations on experimental research typically rely on visual aids. Still, there is room for creativity; for example, you might first walk the audience through the core story without visuals, then retell it with slides that provide vivid supporting evidence. In any case, let us start at the beginning...

Plan the talk. Then prepare the content

Every great talk begins long before you face your audience:

Plan. The first step is to develop a clear, logical plan. Resist the temptation to open presentation software too soon. Instead, sketch ideas on paper or a whiteboard, where they can evolve freely. From the outset, keep your audience in mind: who they are, what they know and what you want them to remember. Let these considerations guide the direction of your talk.

DOI: 10.4324/9781003727521-60

Content. A presentation is not a list of results but a story. Its backbone often mirrors that of a well-structured paper,[2] adapted to the scope of the talk. Begin with the purpose, relevance and need for the study; a compelling introduction draws listeners in. Situate your work within previous studies and the state of the art to give the audience the grounding required to follow your reasoning. Then describe your experimental design: the variables tested, the logic behind any parametric studies or the principles of dimensional analysis that guided your approach. If devices are involved, provide a clear description of their design, size, instrumentation, boundary conditions and noise control. Do not trivialize sample preparation, as it may play a critical role in your results and their reproducibility. Present results to highlight trends and comparisons, followed by a phenomenological interpretation of the underlying processes. Support these insights with mathematical or numerical analyses where appropriate. Acknowledge limitations. Finally, distill your key findings and identify the open questions that remain.

Designing slides. Visuals should support your message, not compete with it. Design each slide with care because details matter. Consider the time needed to present each one clearly while maintaining rhythm. Aim for fewer than 20 words per slide. Include only what you can effectively convey within the allotted time. Animations, flashy transitions and unnecessary decoration rarely add clarity and often distract. Uncluttered visuals help the audience stay focused on your message.

Form matters. Use large, uniform-thickness fonts, as what looks clear on your monitor may be unreadable in a large room. Insert high-resolution figures, cropped carefully, and properly credit every borrowed image, model or idea. Align content along consistent vertical and horizontal lines, maintaining even spacing. Be mindful of the visual "center of gravity", which may not coincide with the geometric center if titles or margins are present. Once all slides are ready, review them for consistent style, including font, color palette and layout. Place page numbers unobtrusively yet visibly in a quiet corner and remove visual clutter such as bullet points, underlining, backgrounds or affiliation logos. Do not let visual noise steal your audience's attention.

Practice, practice.... practice

There is no substitute for practice. Find your voice, focusing on pace, volume, rhythm and tempo. Rehearse until the flow of your talk feels natural and your timing stays under control. Practice out loud, ideally in conditions similar to the actual event, and pay attention to posture. Do short rehearsals of about a quarter of the allotted time to distill the core message you must convey. Prepare so that you can deliver a great talk even if the projector fails and no visuals are available. When rehearsing for a broad audience, *think as wise men do, but speak as the common people do.*[3] Practice until you feel you "got it", when the sequence of ideas becomes second nature without stiff memorization.

Mental readiness

On the final day before delivery, give the presentation your full focus. Set distractions aside and allow yourself to "get in the zone". Cultivate the right mental, physical and emotional state and concentrate on presence and composure.

The night before

Prime your mindset. Practice voice control, from pace and tone to modulation. Stress can impair performance, so plan: imagine what could go wrong and put systems in place to prevent it.[4] Then close your eyes and visualize the moment of delivery: your posture, the flow of your thoughts and your humble yet confident responses to questions. Picture yourself calm and engaging, delivering a great talk. Finally, imagine the moment after it ends and the sense of accomplishment and pride you will feel.

The day of the presentation

Loosen up about six hours before the talk. Go for a brisk walk or a 30-minute run to raise your heart rate and take the edge off. Follow with a long shower to relax. Then review the presentation as a whole, perhaps by imagining you are telling the story to an old friend.

Delivery brings your content to life

Arriving

Dress professionally and arrive early. Punctuality shows respect, and only the most extreme circumstances excuse tardiness. Use the early minutes to position yourself, test the technology and settle in. And yes, make one last trip to the bathroom if needed!

Minutes before your start

Warm up your voice.[5] Prime your posture and focus your mind. Even a brief power pose and confident stance can enhance your nonverbal presence and project confidence.[6] Take two or three deep breaths... here you go!

The talk

Audiences remember the opening and closing best, so prepare both your first and last sentences carefully. Do not rush into a new slide; let the audience absorb it. Speak to the people in the room, not to the screen. Address them one sentence at a time, as if sharing something personal with each individual.

Great presentations have rhythm and tempo that keep the audience engaged; a dynamic back-and-forth between what is and what could be; and a rollercoaster of contrasting flavors, ups and downs, past and future, frustration and hope, analysis and emotion. Science communication is storytelling.

Presentations in "teaching mode", sharing knowledge rather than merely displaying it, are often the most engaging and well-received. Speak with clarity and conciseness; speak from your heart. Strike a balance between assurance and humility: confidence is welcome, but arrogance and showing off are not. Give your best, respect your audience's time and never underestimate them. If the projector fails or anything else goes wrong, resist the urge to complain – *the show must go on*. Finally, always finish within your allotted slot; it is unpleasant to be asked to stop.

Q&A

Listen carefully to each question and try to grasp what is really being asked. Short, clear answers will always resonate more than long, meandering ones. If you do not know the answer, say so honestly and promise to follow up. If someone in the audience becomes difficult, imagine them waiting in line at airport security: just another ordinary person; you may acknowledge their comment, but it is usually better to avoid getting drawn into a debate. Remember, no one in the room knows your work better than you do. Questions reflect the audience's own perspective and naturally pull you toward their frame of reference; your goal is to guide them back and play to your strengths.

Online presentation

Online talks preserve most of the elements discussed above, but many of the cues speakers normally rely on, such as eye contact, body language and subtle feedback from the room, are lost. Instead, one often speaks to a sterile screen, facing muted microphones and disabled cameras. In this setting, clarity, structure and vocal presence become even more important than in person. Speak at a slightly slower pace, articulate more deliberately and use pauses to give the audience time to think. Because technology becomes part of the stage, prepare for its possible failure. Log in early; check your microphone, camera framing and lighting; reduce visual distractions; and choose a neutral background. If technology falters, be ready to restart without complaint and remain focused on the story you are telling rather than on the medium conveying it.

Final advice

Ideas come alive when they are heard, understood and remembered. Preparing a presentation is not just about assembling slides; it is about creating a shared experience with your audience. Watch great speakers and storytellers, such as Martin Luther King, delivering *I Have a Dream*. Be inspired by excellence, like Pavarotti giving his all in *Nessun Dorma*. Let your thoughts flow with precision and elegance, like an Olympic figure skater on ice. Give your very best every time you are granted the privilege of standing before others to present your findings. Enjoy the moment!

Exercise

Take a current research project or a paper you recently read and create a 12-slide presentation.

- Plan the talk: consider your audience, their background and what you want them to remember.
- Design your slides thoughtfully: ensure visuals support your message and are clear and uncluttered.
- Rehearse under realistic conditions: pay attention to modulation, tempo and posture. Repeat until the talk flows naturally.
- Get the core: prepare a 2-minute version that captures only the essential core of your talk.
- Visualize delivery: close your eyes and imagine delivering your talk. Picture your posture, voice, slide transitions and handling questions with composure.
- Record yourself: make a rehearsal video and compare it with your visualization to identify potential improvements.
- Pair up with a knowledgeable colleague: present your talk, then switch roles and record both presentations. Learn from your colleague's version: note different highlights, slide transitions, flow, word choice and style.

Notes

1 M. Twain (1835–1910), American writer.
2 See essay on *Writing: Crafting Meaning One Piece at a Time (Section 6)*.
3 Aristotle (384–322 BCE), Greek philosopher and logician.
4 Daniel Levitin (1957–), American-Canadian psychologist and neuroscientist. "How to stay calm when you know you'll be stressed". TED Talk.
5 Julian Treasure (1958–), British author and communication expert. "How to speak so that people want to listen". TED Talk.
6 Amy Cuddy (1972–), American social psychologist. "Your body language may shape who you are". TED Talk.

SECTION 7

Closing

Essay 55

FREE TO EXPLORE CAUSALITY?

All events seem to be tied together by a relation of cause and effect[1]

Science rests on a central premise: behind every observed phenomenon lies a cause waiting to be uncovered. The search for causality runs deep through intellectual history, from Aristotle's "four causes" and the theological questions of Aquinas to Hume's skepticism, Laplace's determinism and Poincaré's insights on chaos. In the 20th century, quantum mechanics revealed a world that could be described only in terms of probabilities rather than certainties. Einstein resisted this view, remarking that "God does not play dice with the universe". Even religions have struggled with the same paradox: how can free will exist under divine omniscience? Causality remains a fundamental yet unprovable assumption of science. If the universe is truly deterministic, governed entirely by cause and effect, how can we be certain that we are uncovering the right answer, *the truth*?

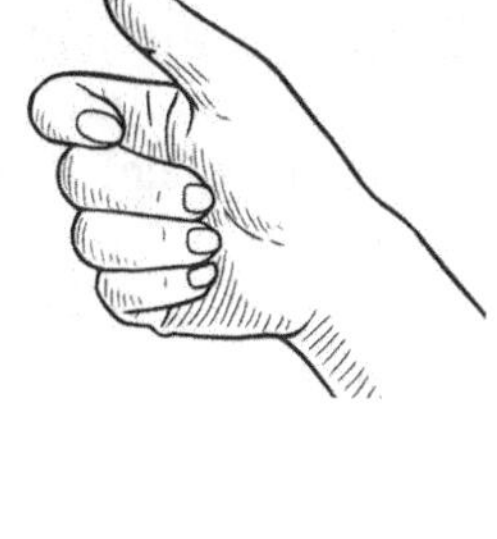

Uncertain abstractions

A coin flip is a familiar metaphor for chance.[2] Yet if we knew every detail of the initial conditions, we could entirely predict the outcome. Reality, however, is both *irreducible* (it cannot be divided without loss) and *incompressible* (it cannot be perfectly captured or predicted). Consequently, all our perceptions and models are abstractions of reality and thus inherently

DOI: 10.4324/9781003727521-62

uncertain. Indeed, if a perfect prediction were possible, the event itself would be its own fastest predictor.[3]

Conceiving randomness - Complexity

Humans are remarkably poor at generating random sequences.[4] Ask someone to write a string of zeros and ones, and patterns emerge almost immediately.[5] Computer algorithms fare little better: most rely on deterministic rules and thus struggle to produce true randomness, an issue with serious implications in fields like cryptology. This raises a deeper question: can genuine randomness emerge from a deterministic system? As complexity increases, the answer seems to shift. Intricate deterministic systems become progressively harder to predict, and randomness becomes more apparent.[6] In such cases, causation does not disappear; it merely hides within complexity.

Nonlinearity - Measurement

Complexity is not always a prerequisite for apparent randomness. Some simple, deterministic systems behave unpredictably because they amplify even the tiniest uncertainties in initial conditions. Moreover, all measurements are inherently uncertain: the act of measuring inevitably interferes with what it seeks to measure.[7] The illusion of randomness thus can arise from the combination of measurement errors and nonlinearities. As J.H. Poincaré (1903) pointed out: "A very small cause that escapes our notice determines a considerable effect that we cannot fail to see, and then we say that the effect is due to chance".[8]

Self-determination

When causation becomes hidden in complexity and uncertainty, a deterministic system can appear to "decide" its own behavior. In such cases, what we call *freedom* may be the unwarranted feeling of self-determination in a reality too intricate to fully decode.

Persistent tension

Clearly, we are pulled by two conflicting forces. One is our belief in *self-determination*, our ability to choose and our creative potential. The other is our commitment to *causality*, the desire to uncover the laws that govern every process... even our own existence!

Exercises

1 Explore the distinction between causality and determinism.
2 Consider uncertainty propagation in nonlinear systems. What are the implications in fields such as economics, climate science, medicine and engineering? And in your own area of research?

Notes

1 D. Hume (1711–1776), Scottish philosopher and historian, key figure of empiricism and skepticism.
2 This sketch is based on a picture sourced from the internet.
3 R.V. Jensen (1987). Classical Chaos. *American Scientist*, pp. 168–181.
4 See essay on *Biases in Decision-Making (Section 2)*.
5 R.M. Hogarth (1980). *Judgment and Choice, The Psychology of Decision*. John Wiley and Sons, New York.
6 K. McKean (1987). The Orderly Pursuit of Pure Disorder. *Discover*, January, pp. 72–81.
7 Heisenberg's Principle of Indeterminacy.
8 J.P. Crutchfield, J.D. Farmer, N.H. Packard and R.S. Shaw (1986). Chaos. *Scientific American*, 255, pp. 46–57.

Essay 56

PATHOLOGICAL SCIENCE AND MISCONDUCT

The community of scientists is bound by a set of values, traditions, and standards that embody honesty, integrity, objectivity and collegiality[1]

Science is the pursuit of knowledge and truth, guided by rigorous methods and a commitment to objectivity. Yet human nature leaves it vulnerable: bias, ambition and error can distort and, in some cases, undermine, even the most careful research.

Integrity in research

Pathological science often stems from overly "wishful" data interpretation. Researchers may unconsciously align findings with expectations or preconceptions, even when evidence is weak or contradictory. Cognitive biases, such as confirmation bias,[2] can make it easy to highlight favorable results while ignoring the rest. In pursuit of a compelling theory, weakly significant findings are sometimes overstated, presented as breakthroughs alongside a "great" theory.

Scientific misconduct, by contrast, involves intentional breaches of ethical norms. Common forms include failing to credit key contributors, adding non-contributing authors, plagiarism,[3] misuse or theft of unpublished or private data,[4] and fabrication or falsification of results. Misconduct can also involve unethical treatment of research subjects (human or animal) or undisclosed conflicts of interest. Integrity failures include selectively deleting data to create a cleaner narrative or misusing statistical methods to exaggerate significance.

DOI: 10.4324/9781003727521-63

Questionable research practices, while often unintentional, can nonetheless undermine trust: failing to retain key data, keeping inadequate records, withholding data from peers, insufficient supervision of junior researchers or releasing unverified results. With powerful AI tools now entering research, scientists face the challenge of balancing gains in efficiency and innovation against the risks of careless or deceptive use - and the potential impact on their own long-term intellectual development.

Case histories

Historical examples highlight human fallibility along the research path:

- G. Mendel (1822–1884) cultivated over 28,000 pea plants and laid the foundation for genetics. Yet reanalysis of the data shows results suspiciously close to expected ratios, suggesting selective reporting or unconscious confirmation bias.
- Polywater (1960s) excited the scientific community with claims of extraordinary properties in water. Later investigations revealed contamination and experimental error.
- Cold fusion (1980s) captured worldwide interest after reports of room-temperature nuclear fusion, but these claims were later debunked and traced to faulty data and misinterpretation.
- Stem cell scandals (2000s) exposed fraudulent data and fabricated results.
- More recently, retractions in oncology have revealed ongoing issues with plagiarism and data falsification.

Together, these cases illustrate the spectrum from unconscious bias to deliberate misconduct.

Do not lose your compass

Pathological science, misconduct and questionable research practices erode trust, the foundation of scientific collaboration. They put colleagues at risk, fuel the "replication crisis" and undermine the credibility of entire fields. Beyond academia, breaches of integrity shake public confidence,[1] from climate science to vaccine research, and threaten to undermine informed policymaking. Vigilance, transparency and humility remain the scientist's strongest safeguards. Question assumptions rigorously, keep meticulous records and share data "as open as possible, as closed as necessary"[5] (always in compliance with legal and institutional requirements), and act ethically.

Exercises

1 Choose one example of pathological science. Examine the study carefully and identify early warning signs that researchers might have overlooked or misinterpreted.
2 The case histories of scientific misconduct listed above are widely discussed in the scientific literature. Review the original publications, revisiting the claims made, the evidence presented at the time and the factors that later led to reassessment or rejection. Summarize each case and note common patterns.
3 Propose concrete steps or practices that ensure adherence to academic and research ethics. Consider supervision, data management, communication and accountability.

Notes

1 The National Academies of Science (1992). *Responsible Science: Ensuring the Integrity of the Research Process: Volume I.*
2 See essay on *Biases in Decision-Making (Section 2).*
3 See essay on *Plagiarism: The Subtle, the Bad and the Ugly (Section 6).*
4 The role of R. Franklin's "Photo 51" in J.D. Watson and F.H.C. Crick's discovery of the DNA double helix is a salient example of how critical data (sometimes shared without consent) can shape scientific breakthroughs.
5 This reflects the FAIR (Findable, Accessible, Interoperable, Reusable) data principles.

Essay 57

FINAL THOUGHTS FOR EXPERIMENTAL RESEARCHERS

Life is a journey, not a destination[1]

Going through these essays has been a journey of discovery for us and, we sincerely hope, for you as well. It is not only about conducting research and sharing the lessons learned from our own experiences and observations but also about understanding ourselves and the life choices that have accompanied this journey. In this closing essay, we reflect on both.

Experimental research: alive and evolving

Experimental research is invigorated, not replaced, by new technologies. Models approximate reality, simulations assume physics, and AI processes vast amounts of data. Yet only experiments confront reality directly: measurement allows comparison, prediction and rigorous understanding; without it, claims remain vague or speculative.

The future experimentalist will be conceptual, strategic and interdisciplinary. AI cannot define context, reframe questions or judge what is meaningful; these remain our strengths. Instead, AI amplifies insight, helping researchers ask deeper questions, explore complex systems and uncover principles previously out of reach. It also highlights gaps in knowledge, pointing to where empirical work is most needed.

Laboratories will grow more instrumented and automated. Experiments will integrate with analysis, simulation, adaptive modeling, digital twins and AI. Machines may optimize protocols and reduce bias, but running an experiment and processing data are not the same as creating knowledge. Insight and judgment remain uniquely human.

DOI: 10.4324/9781003727521-64

Frontiers are expanding. Experiments will probe scales from atoms to planets, and extreme conditions from deep geothermal to nuclear fusion. Natural experiments arising from climate, population and energy transitions will continue to teach us. Medicine, from DNA-centered therapies to neurological research, will increasingly benefit from experimentation guided by AI and rich databases. Beyond the lab, researchers will face questions of human–technology interaction, social inequality and global instability.

Tools and skills

Who thrives in this space? Those who understand processes in depth, who master sensing systems and the logic of algorithms, and who can question observations and challenge the status quo. Curiosity, clarity, skepticism and a careful awareness of uncertainty continue to be the timeless pillars of research.

Know your tools

A researcher's mind is a living workshop. Each method you learn and every physical principle you internalize becomes part of your intellectual toolbox. Like great painters who learned from the masters, hone your skills by observing and analyzing creative experimentalists.

A solid grasp of fundamentals strengthens experimental design, sharpens intuition and helps interpret anomalies when they arise. Do not merely use instruments; learn how they work, how to calibrate them and where their limitations lie. Good experimentalists obsess over design. They refine existing devices to meet their needs and sometimes build their own. This opens possibilities for exploring phenomena no one has seen before. Innovation leads discovery.

Write

Block the best hours of the day for writing. Writing is an integral part of research; it will make you a better researcher. Let writing sharpen the way you think, reason and communicate.

Effortlessness is a myth

Just as athletes and musicians train relentlessly to reach world-class performance, intellectuals must also sweat to excel. Reading about experiments or watching others perform them is no substitute for doing them yourself. Groundbreaking discoveries and elegant writing may look effortless, but behind them lie years of sustained effort. Research rewards those committed to the long haul. Talent may open a door, but discipline, persistence, patience and preparation build a career. Breakthroughs are rarely sudden; luck favors the prepared mind. Build on your strengths. Work hard. You become what you practice every day.

Avoid overcrowded fields: Be different

Choose questions that are meaningful, timely and aligned with scientific and societal needs. Sometimes the boundary of knowledge is foggy, and you may feel you have lost your compass, but what matters most is not losing your purpose. *Seek la différence.* There are singers with modest vocal range who captivate audiences because they sing with their souls; find your own voice. Manage the risks: keep one foot on the foundation you have built, while devoting part of your life to exploring the new. Originality often lives at the intersection of disciplines. Pay attention to what others overlook.

Manage your time

Multiple time scales govern research: gathering information, processing it, identifying key questions, designing tests and interpreting results. Today, access to information is nearly instantaneous, but the volume has created a new bottleneck. Identifying meaningful, impactful and interesting questions is harder than ever. Be patient. Set clear short-term goals and advance steadily.

The experimentalist's mindset

Experimentation, like life, is defined not only by its outcomes, but by the insights gained along the way.

Life as experiment, experimentation as being

To live is our greatest experiment. We explore, adjust and learn within the limits of our abilities and uncertainties. Feedback is often ambiguous, control is partial and chance plays a role. Over time, we realize that our sense of control is limited. One of the deepest outcomes of experimentation is internalizing reality in its complexity. The experimenter is always part of the experiment.

Inner drive matters

Research demands energy and resilience. Purpose drives it,[2] but enjoying the exercise is the most durable source of energy for sustained long-term commitment. Wake up wondering and go to sleep having learned something new every day.

Success has a dual nature

Too much success can distance researchers from hands-on work, while too little recognition can be discouraging. This tension is an inevitable part of a researcher's life and must be navigated with perspective and balance; it is the Yin and Yang of research. As Nietzsche observed, creativity flourishes not in the absence of conflict, but through the productive tension between opposing forces.[3]

Rejection is part of the process

Papers and proposals will be rejected. Experiments will flop. Talks will miss their mark. Keep pushing forward without making excuses for unsatisfying results. History is full of examples where rejection preceded transformation. In 1863, the Salon des Refusés gathered works rejected by the Paris Academy; Manet, Whistler and Cézanne, among others. But what began as an exhibition of "failures" became a turning point in the history of art, revealing new aesthetics that would redefine modern painting. Even the greatest athletes do not win every moment: tennis player R. Federer won only 54% of his points. Learn to absorb setbacks quickly and move forward with clarity. Failure is part of the process; transforming feedback into action is essential.[4]

An unconventional life

The researcher's life can be quite unique: mind-wondering, engaging deeply with meaningful questions and collaborating with thoughtful colleagues around the world. The path forward is often marked by uncertainty and constraint, yet looking back, it can be profoundly rewarding. Reflecting on and writing these essays together has been one such privilege for us.

Your life: Find your balance

A research career is demanding and will test your limits. Keep perspective: do not let setbacks shrink your world, nor let success detach you from what grounds you. Your identity is not defined solely by your research: make space for hobbies, exercise, family and friendships. Some great contributors chose to sacrifice other aspects of their lives for science, often at great personal cost. Learn from their work, but also from their lives, and choose your own path.

The research community

Research begins in the minds of individuals, but grows, becomes real, and endures through human interaction.

Research is a communal life

Progress depends on colleagues, collaborators and mentors. Treat people generously. Share credit. Share ideas. Be courteous. Build trust. Seek mentors who challenge you honestly. Cultivate a respectful relationship with your advisor; it can shape your career. Do not isolate yourself: have lunch with colleagues and visit other labs. Build a global community.

Things not to do

Do not allow yourself to be drawn into toxic dynamics. Avoid meaningless competitions and personal agendas. Do not measure yourself by others' awards or resources. Focus on doing good work and contributing generously.

Final word

Time is our most precious resource: use it wisely. Live meaningfully. Stay curious. Think deeply. Leave the world better than you found it. Enjoy the journey!

Exercises

1 Reflect on a recent setback in your research. How can you reinterpret it as constructive feedback? List two or three alternative paths or strategies to move forward and grow from the experience.
2 Highlight lines in this essay that resonate with you. What do these choices reveal about your values, strengths and aspirations? How could they influence your approach to research, learning and personal growth?
3 Revisit sections of the book that captured your attention or sparked ideas for growth. Summarize what you learned and create a concrete plan to apply these insights in your research, studies or daily life.

Notes

1 Inspired by the writings of R.W. Emerson (1803–1882), American essayist, lecturer and poet.
2 V. Frankl (1905–1997), Austrian psychologist and Holocaust survivor, founded Logotherapy.
3 F. Nietzsche (1844–1900), German philosopher.
4 *Failure has a thousand explanations, but success needs none* is a saying attributed to A. Guinness (1914–2000) and a favorite motto of J.K. Mitchell (1930–2023).

NOTATIONS

a [m/s²]	Acceleration
A [m²]	(Cross-sectional) area
A	Amplitude
b [m]	Thickness
B [m]	Width
B [T]	Magnetic field
B [Pa]	Bulk modulus
c [mol]	Concentration
c [J/kg·K]	Specific heat
c_0 [m/s]	Speed of light
C [F]	Capacitance
d [m]	Diameter or arm length
D [m²/s]	Diffusion coefficient
e	Error
E [J]	Kinetic energy
E [Pa]	Young's modulus
E [V/m]	Electric field
E_x, E_{mix}	Extensive property (component or mixture)
E_y	Error between the mean measured value and the true value
f [s⁻¹]	Frequency
F [N]	Force
g [m/s²]	Gravitational acceleration
G [N·m²/kg²]	Gravitational constant
G [Pa]	Shear modulus
h [m]	Gap, height, size, thread pitch, wall thickness

H [N]	Horizontal force
i	Gradient
i [A]	Electric current
I [cd]	Luminous intensity
I [W/m^2]	Sound intensity
I_x, I_{mix}	Intensive property (component or mixture)
J [mol/m^2/s]	Diffusive flux
k [N/m]	Spring constant
k [m/s]	Hydraulic conductivity
k_T [W/m·K]	Thermal conductivity
ℓ [m]	Length
L [H]	Coil inductance
L	Norm
m [kg]	Mass
M [N m]	Moment
M [Pa]	Constrained modulus
n [–]	Number of tests, variables, or turns
n [–]	Porosity
n [mol]	Amount of substance
N [N]	Normal force
N [–]	Number of teeth, elements in an array, or cycles
p [Pa]	Pressure
P [W]	Power consumption
q [C]	Electric charge
q [m^3/s]	Flow rate (in = inflow; out = outflow)
q_T [W/m^2]	Heat flux
Q [–]	Quality factor
r [m]	Radius or radial distance
r [mol/L·s]	Reaction rate
R [Ω]	Electrical resistance
Ra [m]	Roughness average
S_s [m^{-1}]	Specific surface area
t [s]	Time
T [K]	Temperature
T [s]	Period or characteristic time
T [N·m]	Torque
T [N]	Shear force
u [m]	Particle displacement during wave propagation
v [m/s]	Velocity
v [V]	Voltage
V [m^3]	Volume
W [N]	Weight, axle load
x [m]	Position in space

z [m]	Depth
z [Ω]	Impedance
Z [–]	Atomic number
α [K^{-1}]	Thermal expansion coefficient
α [rad]	Pitch angle
α	Attenuation coefficient
α [s^{-1}]	Rate constant
β [m]	Skin depth
γ [N/m^3]	Unit weight
δ [m]	Deformation, displacement
δ [–]	Impulse
ε [F/m]	Electric permittivity
ε_0 [F/m]	Vacuum permittivity
ε [–]	Strain
ε	Additive error
η [Pa·s]	Viscosity
η	Multiplicative error
θ [rad]	Angle
κ [m^{-1}]	Wave number
κ' [–]	Relative permittivity
λ [m]	Wavelength
λ [–]	Number of levels for a given variable
μ [Pa s]	Liquid viscosity
μ [H/m]	Magnetic permeability
μ [–]	Coefficient of friction
μ	Mean
ν [–]	Poisson's ratio
Π [–]	Dimensionless group
ρ [kg/m^3]	Mass density
ρ [Ω m]	Electrical resistivity
σ	Standard deviation
σ [Pa]	Stress
σ [1/Ω·m]	Electrical conductivity
τ [Pa]	Shear strength
ϕ [rad]	Phase
ω [rad/s]	Angular frequency, velocity
ω_r [s^{-1}]	Resonant frequency

INDEX

Pages in **bold** refer to tables.

For Product Safety Concerns and Information please contact our EU representative GPSR@taylorandfrancis.com
Taylor & Francis Verlag GmbH, Kaufingerstraße 24, 80331 München, Germany

www.ingramcontent.com/pod-product-compliance
Lightning Source LLC
LaVergne TN
LVHW012334130926
847902LV00028B/396

* 9 7 8 1 0 4 1 2 0 3 1 1 7 *